ANALYTIC STUDIES IN
PLANT RESPIRATION

F. F. BLACKMAN

ANALYTIC STUDIES IN
PLANT RESPIRATION

BY THE LATE

F. F. BLACKMAN

CAMBRIDGE
AT THE UNIVERSITY PRESS
1954

CAMBRIDGE UNIVERSITY PRESS
Cambridge, New York, Melbourne, Madrid, Cape Town,
Singapore, São Paulo, Delhi, Mexico City

Cambridge University Press
The Edinburgh Building, Cambridge CB2 8RU, UK

Published in the United States of America by Cambridge University Press, New York

www.cambridge.org
Information on this title: www.cambridge.org/9781107619487

First published 1954
First paperback edition 2013

A catalogue record for this publication is available from the British Library

ISBN 978-1-107-61948-7 Paperback

CONTENTS

PREFACE

In 1936, at the age of 70, F. F. Blackman retired from his position as Reader in Botany in the University. During his retirement he hoped that, with the secretarial help of one of his research students, he would be able to prepare for publication much of the work on which he had been engaged. But the Second World War came and his hopes were not to be realized. Until January 1947 he had been continuously occupied with this work, and on his death left much in varying stages of completion; some had been made nearly ready for publication many years before. The carefully documented editions of some of the papers make clear, what was already known to his friends, that Blackman was a man not easily satisfied. Now that he had gone, ought these papers (which apparently had not satisfied him) to be left unpublished; or ought a wider circle than his students to have the privilege of seeing how his mind worked? There was a series of six papers dealing with the effect of varying oxygen concentrations on the respiration of the apple, to which he had given much thought, and as these could be prepared for publication without material alteration to the typescript, it was decided, with Mrs Blackman's consent, to publish these. To make the account more nearly complete it has been necessary to include three papers already published by the Royal Society, and for permission to reprint these we have to thank the Council.

These nine papers show Blackman's ideas developing. The incompleted drafts of further work leave little doubt that before he died his ideas had developed yet further, and that he would not have subscribed to some of the views published now.

The experimental work on which these writings of Blackman are founded was done by one of his research students, P. Parija. One of the published papers bears his name alone, another his name in conjunction with Blackman's; Parija's name also appears on one of the typescripts, but it is quite clear that all these papers show the work of Blackman's mind, and that he himself wrote the papers, and the book therefore bears Blackman's name alone; information about names appearing in these papers is supplied in editorial footnotes. Professor Parija has approved this procedure. It did not seem necessary to distinguish the editorial footnotes from the others.

Having explained the origin of this book it is necessary for us to say something more about the writer and its contents.

Blackman was never content to publish unanalysed data. His interest was in metabolic concepts, and any set of data was analysed exhaustively to establish and interpret the underlying correlations. In this way Blackman subjected to his own special methods of analysis many series of results; some of these were obtained in researches carried out under his supervision, but others were from published results of other workers. Little of this careful analysis reached the stage of publication; nevertheless, these analytical studies of experimental data comprised the bulk of his lectures and gave them their unique character.

The nine papers published here are based almost entirely on the data collected in one year (1920–1) with twenty-one apples. Under his guidance research students investigated other aspects of apple respiration as is shown in Appendix II. The results would have formed the subject of other contributions, but no doubt they influenced his lines of thought as he wrote the present papers. And this is probably true also of investigations made by others of his students on the respiration of other plant material.

The six unpublished papers in this book were written mainly between 1930 and 1938, and it might be thought that the material is now out of date. This is not so. The experimental data here printed are still among the most important material in this particular field, and Blackman's analysis of the data in these papers provides the most comprehensive picture yet available of the influence of oxygen on respiration. Blackman's initial analysis of one section of the data, published in the three papers of 1928, was recently referred to as follows: 'The classical paper of 1928 remains far and away the most careful and critical analysis of respiration data in the field of plant physiology.'

In the three published papers Blackman was considering mainly the effect of nitrogen and of air on the respiration of apples. In the six additional papers he is concerned with the influence of the full range of oxygen supply, from zero up to 100 % oxygen. Many physiologists have investigated the influence of nitrogen and of air on respiration, but few have realized the importance of studying the effect of the intermediate concentrations of oxygen, and it was the results obtained with these that Blackman found so revealing. He writes in Paper V: 'We may present fig. 4*a* as the first comprehensive schema that has been put forward, for a higher plant, for the effects of oxygen supply upon the full range of respiratory metabolism between the points of the air supply and the zero supply in nitrogen.'

Fig. 4*a*, just mentioned, shows the changes in the relative rate of CO_2 production in relation to oxygen supply. As such it represents only experimental facts and provides no metabolic concepts. Blackman said of it in Paper VII: 'This set of curves was slowly built up from a multitude of observations by the empirical procedure described in Papers IV and V. We have now to...carry out upon the set an analytic procedure which aims at disentangling the functional determinants that lie behind its graphic features.' From this analytic procedure as carried out in Papers VII and VIII emerges fig. 10 of Paper VIII, in which is put forward schematically a new concept, namely, that in plants a main influence of oxygen concentration on respiration is through an effect on the rate of supply of carbohydrate substrate; to this process Blackman gave the term carbactivation. The importance of this aspect of oxygen supply was indicated in the published Paper III in the following words: 'Should it be established that the primary effect of varying oxygen supply in respiration lies in the control of carbohydrate equilibrium, then our biological outlook on this function will be considerably modified.'

But although there is in the papers more than one important metabolic concept which will provide a stimulus for further research, the scientific contribution which the series makes may be of less ultimate importance than their influence on the analysis and presentation of physiological data. Each of the nine papers is remarkable for the scrupulously careful arrangement of the sequence of ideas and the clarity of the writing. Scrutiny of Papers VII and VIII will reveal the stages in Blackman's analytical approach. First came an exhaustive graphical analysis of the experimental data, to establish physiological relations; this was followed by careful consideration of the existing theory, the outcome being that the theory was stated in its simplest form, and certain independent physiological determinants were selected. By graphical analysis the influence of variations in the independent determinants was then investigated, and in this manner it became possible to compare theoretical schemes with the experimentally observed relations. Finally, the analysis led to the presentation of a simple schema indicating the physiological systems involved. Much of Blackman's approach is summarized in the following phrase, which is taken, not from his papers, but from a cutting found with his notes—'the compulsion of theoretic possibilities pushed to their logical conclusion'.*

To some Blackman may seem to be too dogmatic; to others this characteristic will indicate his search for simplicity and clarity. To quote from Paper IX: 'Though, in order to economize words these views are expressed as antithetically as possible, it is not intended to give dogmatic support to the view that has been developed out of observations on plants. The intention...is rather to formulate the opposition of views as clearly as possible so that further data may be collected to clear up the critical issues between the theories.' And from Paper III: 'After that it may become necessary to take the present schema to pieces and reconstruct it, but at least we shall have consolidated a mass of relations to which any future system must conform.'

The papers were not prepared as chapters of a book. There are repetitions, and in some places the treatment in later papers replaces that in early. To try to make a unity would involve risks that we were not prepared to take. By leaving the papers essentially as they were we have left it possible for a reader to trace the development of the author's ideas.

Although the individual papers and their different editions had been prepared by Blackman with ultimate publication in mind, their assembling has involved Dr J. Barker, one of Blackman's pupils, in much devoted labour.

Appendix II contains a list of the names of Blackman's research students, as complete as we have been able to make it. Where the results of the investigation were embodied in a dissertation submitted for a research degree, the title has been given. Some of the results have already appeared as published papers.

* Bukharin, *The Times*, 8 March 1938.

G. E. BRIGGS

Cambridge
1952

I

THE RESPIRATION OF A POPULATION OF SENESCENT RIPENING APPLES*

CONTENTS

INTRODUCTION

Of all protoplasmic functions, the one which is, by tradition, most closely linked with our conception of vitality is the function for which the name of respiration has been accepted. It might, therefore, well be expected that every variation of the intensity of the metabolic activity of a cell would be correlated with some change in the respiration of that cell. Before we can decide whether respiration really holds this position as an index of the integrated activities of the cell, we need to accumulate respiration data for different types of plant organs throughout their life history of development, maturity and senescence. These data must then be examined critically with the hope of establishing the nature of the major and minor determinants of the variations of intensity of respiration. No such collection of data has yet been published. The present paper aims at making a contribution to this collection and other contributions should follow.

A good deal of work has already been carried out in the Cambridge Botany School upon respiration of evergreen leaves, which continue to exist in a state of maturity for very long periods of time. In striking contrast with this type of organ is the type of the ripening fruit. The natural biology of the two is so different that it is obviously of importance to establish whether the same fundamental principles are manifest in both. In the ripening fleshy fruit, senescence is the dominant stage of ontogeny. The fruit of the apple, which possesses such striking keeping properties, is most suitable for investigation, since it runs through its ripening senescence at a slow rate.

An opportunity of taking up such work was provided by the beginning of the experimental work of the Food Investigation Board, at Cambridge, under Sir William Hardy. The apples made available for us were those that were kept in cool storage at about 2·5° C. for the investigations of Dr F. Kidd and Dr C. West.

* This paper was published in 1928 as the first paper in the series 'Analytic Studies in Plant Respiration'. The authors were F. F. Blackman and P. Parija, the reference being *Proc. Roy. Soc.* B, **103**, 412, 1928.

We are indebted to the Board for a subsidy to enable the junior author to devote a year to the investigation presented in this and the following papers.

The outlook and the problems. Our outlook upon these apples has been to regard them as a population of individuals slowly progressing in cool storage through the metabolic drift which constitutes the senescent and penultimate stage of ontogeny, popularly spoken of as ripening. Some biological truths of this drift can be brought out by statistical treatment of the population as a whole, others only by intensive study of individual behaviour. It is the latter aspect that we wished to explore, in order that we might find out which features of individual respiration can be held to be capable of physiological interpretation and which must, at present, be regarded as indeterminable chance happenings.

The nature of the test applied to the respiration of the apple population was to take out of store, throughout a period of eight months, individual apples, one by one, and examine the intensity and course of their respiration in air; and also, as part of the same inquiry, to subject them to a variety of oxygen mixtures ranging from zero to 100 %. The survey of the results in air is given in the present paper, while those in the oxygen mixtures are brought together in the following papers.

The previous history of the population was that they were Bramley's Seedling apples grown on fen soil, gathered at the beginning of October 1920, and maintained in cool storage between 2 and 2·5° C. for the investigations of Dr Kidd and Dr West, to whom our thanks are due for this essential assistance. The apples were picked from one orchard at one time and believed to be a homogeneous population, though they had not been gathered or graded under scientific supervision. When once brought into store the population is, of course, exposed to an extraordinary and quite unnatural uniformity of environment—no change of temperature or humidity, and no alternation of light and dark, for months in succession. One interest of our investigation would be to find out how far the population declared itself homogeneous under our physiological tests.

The individual apples were brought to the laboratory under standardized conditions and investigated at one temperature only, namely, 22° C. No conscious selection was exercised in taking individuals from store, except the avoidance of any that were bruised or showed traces of brownness. The average condition of the population was, of course, changing with the progress of the metabolic drift, and this revealed itself by the gradual colour change from full green through yellow-green to golden yellow and finally brown.

Allowing for all this drift, the conclusion was yet forced upon us by the results of our work that the population could not be described as homogeneous. Clearly, the apples continued to be distinguishable by physiological characteristics that differentiated them on the day that they were picked and put into store. The fact that recent differences of environment were negligible as a contributory factor to the observed differences of behaviour encouraged us to persevere in the endeavour to explain observed differences in terms of initial inherent qualities and temporal physiological drift.

Our observations of their respiration at 22° C. were continuous and revealed many minutiae of difference in behaviour, all of which we have endeavoured to bring to account, and either interpret them or formulate problems as to their determination. It is at least made clear, that had three or four apples instead of one been employed in each experiment, yielding merely average results of behaviour, then it would not have been possible to push our analysis very far. It is an essential consequence of the metabolic drift in storage that results obtained in one month are not repeated exactly in the next. When any problem arises in this type of work, it is not possible to go back and repeat an observation on identical material once more.

Experimental methods and procedure. The apples were brought from the cool store to the laboratory, weighed, and placed singly in a glass respiration chamber of a spherical form, which consisted of two hemispherical domes with a wide equatorial flange and two polar tubes as inlet and outlet for the constant current of gas, maintained by aspirators through the chambers, at a rate of about 1500 c.c. per hr. The chamber halves were waxed together, fixed in a weighted frame and lowered into a large thermostat bath kept at 22° C. The whole of this occupied about 30 min.; the air currents were then started at their proper rate and run for about 90 min. as a preliminary before estimations were started. There is therefore about 2 hr. of respiration in all before the point of time that figures as zero time in the records. The bath temperature generally kept constant within 0·2° C.; there were a few misadventures during the nine months' work, which are noted in the records of the individual cases. The current of air passes from the chambers through Pettenkofer tubes of standardized baryta, which are arranged as a parallel set, and the current is shifted on automatically by clockwork from one tube to the next at intervals of 3 hr. In this way continuous records of the production of CO_2 can be obtained for an indefinite period of time. In the present work some records continue for 16 days without a break.

Each day the used Pettenkofer tubes are lifted out one by one, washed into a beaker and titrated with decinormal HCl and phenolphthalein. They are then refilled and replaced in their frame, ready for the air current to come round again. The CO_2 production is expressed in mg. CO_2 per 300 g.hr. Medium-sized apples were selected from the store, averaging 140 g.; they were weighed again at the end of the experiment; the loss of weight averaged 1·5 % per 10 days, the minimum being 0·77 % and the maximum 2·09 %. The individual cases will be found detailed in Appendix I.*

The respiration records. In all, twenty-one experiments, numbered V–XXV, were carried out, in that sequence, on single apples brought from cool store, from the middle of November 1920 to the end of June 1921. The CO_2 production of some was examined in air only, but most were exposed as well to the effects of one or other of the following concentrations of oxygen: 0 (nitrogen), 3, 5, 7, 9 and 100% O_2.

The work involved nearly 2000 estimates of the CO_2 of respiration; the numbers are not tabulated in this paper but presented in graphic form throughout. The graphic records of the respiration values of the twenty-one experiments will be found set out in Appendix I*; these records will be referred to as the 'General

---

* Appendix I will be found at the end of this book; it was originally published as an appendix to 'Analytic Studies in Plant Respiration. II' (*Proc. Roy. Soc.* B, **103**, 446, 1928).

Charts' of the results. Mostly two experiments were carried on concurrently; and where this was so, the two records are grouped together in the same chart, one often serving as control to the other.

In the various sections of this paper, discussing special points, excerpts from the general charts bearing on the problem are brought together and correlated. In all the charts the ordinates express mg. CO_2 per 300 g.hr. for the fresh weight of the apple when taken out of cool storage. The abscissae are hours of time from the beginning of the respiration measurements. Each 3 hr. measurement is represented graphically by a single heavy line, or by three consecutive dots, covering the period of 3 hr. duration.

The long series of continuous estimations show a constant tendency to fluctuate up and down. We have thought that the general drift of the respiration is brought out more clearly in our graphic records when we represent it, not by a single median line, but by two 'contour lines' which are drawn parallel, one above and the other below the range of the fluctuation. When a definite numerical value is needed for respiration, it is, of course, the value midway between the contour lines that is adopted.

The fluctuations. It is not to be expected that under constant conditions the sequence of estimations would give values lying on one steady line, but it is clear that the fluctuations that actually occur are much greater than those that can be attributed to small random errors of titration, tube-washing and manipulation.

As a striking example of these fluctuations, it has several times been noted that, when the respiration is undoubtedly declining generally, as proved by a record lasting several days, there may yet occur in the course of it a level sequence of no less than four identical readings—covering 12 hr.—before the falling drift comes into evidence again. Had the record been stopped just at the end of this 12 hr. it might have been concluded, confidently, that the fall had passed into a definite level phase.

The range of the fluctuations indicated by the distance apart of the two 'contour lines' is about the same in the different experiments when respiration is running an approximately level course, and amounts to 0·7 mg. CO_2, but occasionally the readings seem to swing with greater amplitude. Apple VIII provides a unique case; this was the one apple that developed a patch of fungus mycelium, involving a big rise in the respiration. Here the fluctuations became very great and the successive readings were most irregular, which we attribute to the irregular growth and activity of the fungus on the apple tissue.

The 'air-line'. When some partial pressure of oxygen, other than that of air, is given it will be seen in the various records that the CO_2 production may be either much increased or much depressed, causing a deflexion of the double contour line which indicates the drift of respiration. At such times it is important to know the ratio of this increase or decrease to the magnitude of respiration that would have occurred had the apple been kept in air all the time. For this purpose it is necessary to join together the air records before and after by an interpolation. This is represented in the records as a single median line, and not by contour lines. Often these interpolations have to be long, and much study of the records has been required to carry them out with confidence.

Joining up the actual air records by interpolations, and sometimes adding extrapolations, we can get a continuous line, which we shall call the 'air-line', running right through the experiment and suitable for comparisons and controls.

Respiration in air: the special problems. We may conclude this introduction by indicating the chief features and problems presented by the respiration of our twenty-one apples, when we came to review the records obtained. The earliest apples, V and VI, were investigated when they had been in store for only 30 days; the latest, XXIV and XXV, when they had been stored for 260 days.

(*a*) The primary variable was the absolute magnitude of the respiration for different apples. Some air-lines started as high as 20 mg. CO_2 per 300 g.hr. apple, while others were as low as 12 mg. CO_2. To some extent the drift of these initial magnitudes was temporal, but clearly some other quite different factor was involved as well. This complication is to be examined in the first section.

(*b*) Apart from differences of pitch the air-line records were not all of the same type. Some declined fairly fast, some kept level for a time and then fell, while others rose day after day. The outstanding complication was that these different forms did not present themselves as one uniform drift of type, as the individuals were examined month after month. The resolution of these complications in the form of the drift of the air values is the subject of the second and third sections.

(*c*) A minor feature that attracted attention was the form of the air-line initially, immediately after heating up from 2·5 to 22° C. The rising respiration did not simply mount up to the air-line value but in many cases clearly overshot that value and then fell back to it. This special disturbance of initial rates is examined in §§ 4 and 5 as the 'change of temperature effect'.

1. THE INTENSITY OF THE RESPIRATION OF APPLES DURING THE SENESCENT PHASE

The problem that we have to take up in this section is that of the great variation of intensity of respiration shown by apples removed at intervals from the cool store over a period of 8 months. We have twenty-one cases to consider, and the ideal values to work on would be the initial air-line value for each case at 22° C. Brought, as they are, from 2·5° C. the respiration rises rapidly at first but presently settles down to proceed along the air-line. Extrapolation of this air-line back to zero hour would give, for each apple, what we may call its ideal initial respiration value. There are certain complexities about the early course of observed respiration, which are to be explored in § 4, and these affect the estimation of such initial values; but fortunately the divergences of the apples one from another are so great that for the present section it is a matter of indifference how the initials are arrived at, so long as the same method is followed for every apple. We will therefore adopt the ideal initial values just mentioned, which will be found set out in full in column 7 of the table in Appendix I. These twenty-one initials are represented in fig. 1, plotted against a time axis which gives the date when each apple was removed from store and the number of days that it had existed in store since picking. Clearly the early apples give respiration values that are lowest of all, after which come slightly higher values. Subsequently there is great divergence of values, the high values mounting up to over 20 mg. in March and falling off somewhat towards June. But all through this drift there are occasional occurrences of medium and quite low values. In the figure the highest values have been connected up with one line and the lowest with another, while a median line has been carried throughout the assembly. On this unanalysed presentation of the data one might conclude that there was a marked tendency for the mean value to rise till about March and then to decline somewhat to June. But more marked than the

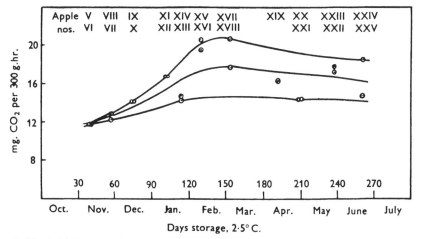

Fig. 1. The initial respiration values at $22°$ C. arranged in chronological series. The time axis gives dates of removal from cool storage at $2·5°$. Twenty-one apples were examined at eleven dates. The exact dates and respiration values will be found in columns 2 and 7 of the table in Appendix I. When two apples at one date gave identical values, this is indicated by the point having a dumbbell surround instead of a circle. The serial numbers of the apples are given by the roman numerals along the top of the figure; the top numeral refers to the apple having the higher respiration value. Lines are drawn connecting up the drift of the highest values, and the drift of the lowest values. A mean line is added equidistant from the two extreme lines.

drift of the mean would be the enormous increase of the scatter about the mean with progressing ripeness. Another curious feature would be the number of examples that lie on the extreme lines and the few that are found on the median line, so that the whole assembly does not at all resemble one showing a normal scatter of chance divergences about a slowly drifting mean. Could apples really diverge so much from one another by unanalysable chance variations, then nothing would be gained by working with single individual apples.

Our first progress in the analysis of this complexity came from comparing with it the grades of ripeness indicated by the colour of the individual apples at the dates when their initial respirations were determined. As the months from November to June passed there were of course changes in the appearance of the apples in store. At first, from October to March, all were full green, but then the slow progression towards ripeness caused visible change of surface colour, through yellow-green, to full yellow and then on to partial or complete brownness. While the whole population of the store was drifting through this series of changes it was clear that all individuals were not moving at anything like the same rate. In April apples representing all stages from green to brown were present. As time went on, more and more apples had to be set aside as brown in parts. Most of the apples came to ripeness (yellow colour) in April to May, but some were still yellow-green at the end of June, by which time no pure green apples were left.

The correlation of colour with initial respiratory magnitude for the last seven cases first supplied a clue to the system underlying the irrational distribution of values. The earlier cases had led one to associate very low initial respiration with unripeness, and high initial respiration with ripening, but when apples XXIV and XXV were selected as the two most extreme apples for unripeness and ripeness, respectively, to be found in the store at that date, XXV being 'golden yellow' and XXIV only just beyond full green, namely, 'yellow-green', then it was found that XXV had a low initial respiration value, 14·7, while XXIV was high, 18·5. This distinction was supported by the preceding apples, for XIX, XX and XXI, with low respiration, had been recorded as yellow, but XXII and XXIII, with higher values, as 'green-yellow'. Thus low initial respiration is associated with unripeness in November to December and with very full ripeness in May to June. We must conclude then that respiration first rises and then falls, and can be clearly associated with the ripening drift of colour in storage at 2·5° C.

The relation between colour and intensity of respiration (as measured here initially at 22° C.) that came out of a close study of their parallel drifts may be formulated somewhat as follows. Every apple picked unripe drifts during ripening through a special senescent phase of metabolism, the essential nature of which will be discussed in § 3. The passage into this phase from the previous phase of metabolic maturity is marked by a rise in respiration rate, which rise starts slowly, progresses faster, and then slackens off to a maximum value; during the early part of this rise the apple colour is full green, losing its intensity towards the maximum of respiration. After the maximum, the respiration begins to fall, though at first slowly, and during this stage the colour of the apple may be described as typically yellow-green. This stage is succeeded by a quicker fall of respiration and the apple is now full yellow colour. This fall of respiration continues, provided no fungus attack develops, on into the stage when the apple becomes brown. In fig. 2 we present these relations schematically as a time drift of the two characters. We do not predicate a very close correlation of colour and respiration, since the former is determined only by surface cells while the latter is an expression of the whole mass of the apple, but the figure gives the relation which seems to be typical.

Observation of a population of apples in storage teaches us at once that individual apples run through this typical drift at different rates, since some may have drifted right through to brownness when others have only reached the green-yellow stage. The implication of this is that the respiration curves of such contrasted individual apples, plotted on the same time axis, will cross one another, for the quick-ripening apple will show an early maximum and an early fall, while the slow-rising respiration of the slow-ripening apple will cut across the fall of the other on its progress to its own later maximum. We are inclined to think that the later the maximum is attained the lower is its pitch for a given apple, as compared with the earlier maximal values of the quickly ripening apple individuals. The evidence points to all the apples having much the same low respiration values in the 'mature' stage before the senescent rise sets in.

In conformity with these propositions we have fitted to our observed assembly of initial values of respiration sets of curves representing the complete respiration sequence for the whole of the senescent drift. This is carried out in fig. 3. Four of the initial points are marked by squares, and these will be dealt with later. At present we are concerned to schematize the seventeen cases represented by circles.

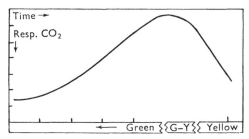

Fig. 2. Schematic form of drift, with time, of intensity of respiration of a ripening apple passing through the senescent phase. The form of curve given here is applicable to the respiration measured initially at 22° C. when the senescence is progressing in cool storage at 2·5° C. In this generalized curve no definite values are given to either ordinate or abscissa axis, but the colour sequence which is associated with the respiration drift is indicated. Specific cases are to be dealt with in fig. 3.

Through these are drawn two sets of illustrative curves, and these curves imply that for any apple in the schema, the curve which passes through its observed initial value at a certain date indicates what its initial respiration would have been, had it been withdrawn from store at any other date, either earlier or later, than it was actually taken out. One striking feature of this figure is the strong suggestion of heterogeneity in the population shown by the fact that the sets of curves fall into two remote groups. Variation could easily be made in the course of the construction curves, but the absence of individuals of intermediate rates of ripening, both in the rising and the falling stages, could hardly be eliminated by an alternative formulation. We propose to distinguish the twelve apples that ripen quickly as representatives of class A, while the five that ripen more slowly may form class B. Below the respiration drifts are set out the colour sequence expected for class A and for class B according to the principles already enunciated. These sequences are based on the recorded colours as given in column 3 of the table in Appendix I.

According to this schema the scatter round the mean is very small for class B, but only a few of this class chanced to be drawn from the population, presumably because they were in small minority. Even in class A the scatter is not great, as the drift lines are here presented. The fact that so often two apples drawn from store at one date give nearly identical respiration values in class A, points also to this being a real group of small scatter, rather than part of one wide common group with the remote cases of class B.*

There are still four other apples that have not been brought to account in our treatment of class A and class B in fig. 3. These are VII, VIII, XXII and XXIV,

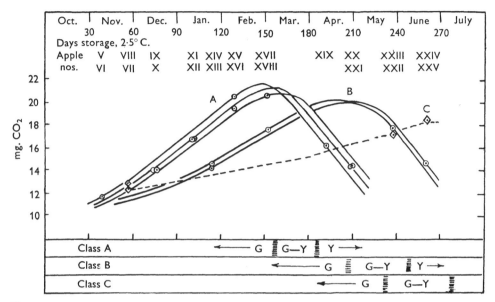

Fig. 3. The observed initial respiration values of the twenty-one apples given in fig. 1 are here grouped into three separate classes A–B–C which ripen and pass through their respiration drift at three separate rates. Through each initial point has been drawn a curve of the type of fig. 2. Twelve apples are allotted to class A, and three typical curves serve to indicate their senescent drift. Five apples are allotted to the later ripening class B, and two curves serve to connect them together. The four apples of class C are connected by a single drift line, which had not run beyond the rising phase when the investigation was ended. The identification numbers that were given to the individual apples are entered in the upper part of the figure over the individual respiration values. Where the two apples at one date have different respiration values the top number is that of the apple with the higher respiration value. The observed colour sequence noted for the apples of each class is set out below to show that the drift from green through green-yellow to yellow takes place at a different rate for each class, and that the relation of these rates for the three classes is the same as that shown by the respiration rates.

represented by squares instead of circles. Apple XXIV presented us with a very high respiration very late in the chronology, and also showed a yellow-green colour. These are features of an apple near its maximal senescent respiration, which suggests that XXIV must represent a class that ripens still much later than

* In addition to a scatter of rates of ripening within class A there must also be some variation between individuals with regard to the respiration value per unit of fresh weight. If this had a considerable range, the vertical divergence of lines within the nest of curves might be wholly or partly an expression of such a scatter. Introduction of this consideration might remove the intersection of curves inside each class at the peak of the schema and substitute a set of three parallel lines, but this would not affect any arguments based on the schema. During storage, month by month, the apples are losing water, so that from this cause alone the respiration values per unit fresh weight must rise. The observed water loss is, however, not enough to make an appreciable contribution towards explanation of the observed large rise.

class B. This apple provides the first individual for our new class C, and to this are assigned also a pair of early apples—VII and VIII. On any evidence that this chart can provide this last attribution is, of course, absolutely arbitrary, for these two apples are perfectly situated for class A apples. But on evidence provided in § 2 on air-line drifts, and confirmed in the next paper where the behaviour in nitrogen is investigated, there is no doubt whatever that VII and VIII must be segregated from their neighbours and classed with XXIV as representatives of a separate class, C. The straight line drawn in fig. 3 from VII to XXIV would serve for the slowly rising limb of the schema of initial respiration values of this class. This line passes through apple XXII, which is also undoubtedly of class C on similar evidence to be set out later.

Such an analytic schema of three sets of lines provides some interesting situations when an apple is found at the intersection of two lines. Thus XIX might be claimed, as far as position on the chart goes, as either rising C or falling A; but the fact that it was full yellow settles it as A. Again, XXII and XXIII are in the chart so balanced between rising C and falling B that we must seek other evidence. A falling B apple, not far below the maximum, should be green-yellow as was XXIII. It would then have been expected that XXII, which we have referred to class C, should have been more green than XXIII. It was not recorded at the time as more green, but only as yellow-green, though a special note was made that XXII was strikingly turgid and fresh in appearance for an apple at that late date, so that on the whole its condition supports its attribution to a rising line.

It may be mentioned that we had no schema of this type before us when the experiments were actually made, but only a growing perplexity about the association of low respiration with both the greenest and the yellowest apples. It was this perplexity that led us to select for the two apples of the late June experiment the greenest apple and the yellowest apple that could be found in the population. This gave a clue which, followed up, has led us ultimately to substitute for the perplexing configuration of fig. 1 the highly rationalized formulation of fig. 3, in which, whether rightly or wrongly, each point finds its place in one of three physiological classes, and also a definite position in the sequence of development of its own class. This formulation on the basis of the evidence so far produced may appear rather unsubstantial, but it will receive further support in later sections on the air-lines as well as from nitrogen effects.

It may have puzzled the reader that, whereas the whole of this section is expounded as a study of senescent drift of a population of apples stored at the temperature of 2·5° C., yet all the respiration values brought to account are for the high temperature of 22° C. The explanation of this indirect approach is that the experimental work was undertaken as a study of the effect of oxygen concentration upon apples at 22° C., and not till after it was finished was it discovered that the data supplied material for the exposition of the various analytic treatments set out in the sections of this paper. The respiration values in the present section are, however, all initial values, at 22° C., and so are determined by the physiological state of the apple at 2° C. when removed from store multiplied by the factor which gives the proper ratio for increase of the respiration rate between

2 and 22°. We have not carried out any respiration measurements at temperatures below 22° C., but the examination of such apples at various low temperatures by Drs Kidd and West suggests that the temperature coefficient, $Q_{20°}$, be given a value of about 8·0.

The really remarkable fact that stands out clearly, in whatever way the data are handled, is that the respiration of an isolated starving organ, at a certain stage of its drift, starts to rise considerably. We may postpone our interpretation of this phenomenon to § 3, and take up in the next section the analysis of the forms of drift of the air-lines of our population of apples.

2. THE COURSE OF THE RESPIRATION AIR-LINE OF INDIVIDUAL APPLES

The next aspect of the respiration of apples in air at 22° C., after cold storage at 2·5° C., that we have to investigate is the general trend of the air-lines for the twenty-one individual apples examined, as their respiration is followed hour after hour for days. In the first section we analysed the phenomena presented by the initial respiratory values, which were found to vary from 12 to 20·6 mg. CO_2, and we put forward a schema which introduced orderly sequences into the apparent disorder of the occurrence of the different initial values. The main conclusion was that the apples must first be sorted into representatives of some three physiological classes, A–B–C, which are characterized physiologically by ripening quickly, intermediately or slowly under the conditions of storage.

We have now to characterize and classify the different courses that the respiration of the apples run after these initial values, and see what physiological order can be introduced into this aspect also. These courses may be, for a long time, either falling, level, or rising, and we have to determine whether, for example, it is the high initial values that are associated with subsequent steep fall, while the low ones keep level, or whether the apples on the ascending limb of the A,B,C schema in fig. 3, p. 9, rise and those on the descending limb fall; or whether perhaps apples of class C behave in a different way from class A, and so on. For this purpose the different types of course run in air must be first brought together for an empirical comparison of forms. This is done in fig. 4. The continuous parts of the lines in the figure represent those parts of the course in which the apple was actually respiring in air, the dotted parts indicate the periods in which the apple was in other gas mixtures than air; taken together these represent the course that the respiration would have followed had the apple been kept continuously in air. The whole composite line, made up of direct observations, interpolations and some extrapolations we speak of as the 'air-line' of that apple. The derivation of these air-lines may be seen by consulting the full records of the experiments given in Appendix I.

In fig. 4 the air-lines are arranged for comparison of forms, one over the other, as close as may be without overlapping, in a sequence that ignores ordinate values and is primarily chronological. The time axis below represents hours of respiration at 22° C. after removal from storage at 2·5° C. The respiration values which we

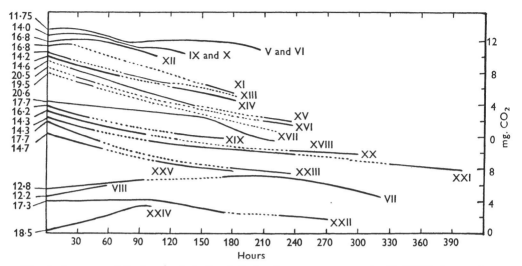

Fig. 4. A survey of the forms of air-line drift for the twenty-one apples V–XXV, the abscissa axis being hours at 22° C. after removal from cool storage. The individual air-lines are not spaced out in their real relations to a single-ordinate axis, but are brought close together for comparison of forms (see the table below). The ordinate scale is indicated on the right and the initial value of each air-line is outset to the left.

Classification of air-line forms

Empirical types based on form	Subtypes	Apples		Fall in first 100 hr. mg. CO_2	
I. Air-lines rising initially; later declining with a steepening fall	I a, fall very long delayed	VII		Nil	
		VIII		Nil	
	I b, fall after moderate time	XXII		Nil	
		XXIV		Nil	
II. Falling air-line of composite course. Fall not more steep at first; form tending to be concave below from steepening fall	II a, course level for an initial period, then transition to fall	V	VI	1·25	1·25
		IX	X	1·5	1·5
		XII		1·7	
		XI		[3·2]	
	II b, approximately rectilinear downward course from beginning	XIII	XIV	3·7	3·3
		XV	XVI	4·3	3·3
		XVII		3·5	
		XVIII		1·3	
III. Falling air-lines; regular continuous curves, steeper at first, slackening later	III a, initial fall moderate rate	XIX		3·2	
		XX		3·1	
		XXI		3·2	
	III b, initial fall somewhat steeper than III a	XXIII		4·0	
		XXV		3·6	

have called the ideal initial values are noted at the beginning of each air-line. After studying these forms, from every point of view, it seemed that we could distinguish three main types, each possibly divisible into two subtypes. These are set out in tabular form and a descriptive label has been given to each type.

Our next inquiry is then to see what relation these types have to the schematic relation of initial respiratory magnitudes established in § 1.

It will be seen that four of the records alone show an upward tendency, and these have been grouped at the lower part of the figure. They are characterized as type I, with the specific character that after 100 hr. of respiration the value is no lower than it was initially. Ultimately these air-lines start to turn down to lower values. The rest of the air-lines have a downward tendency and are to be spoken of as falling air-lines. Among these we can distinguish two contrasted types of fall: there is one type, III, which has a very regular and characteristic form, in that it starts from the beginning with a steepish fall and proceeds perfectly smoothly, falling less and less fast, so as to present a convexity below and ending, if observed long enough, in a practically rectilinear downward slope. The standard of this is record XXI; four others closely conform to it making up the content of type III. The remaining falling air-lines, constituting type II, are less homogeneous and more difficult to define. These are grouped at the top of the figure. None of them starts with a steeper fall than it shows later, and some of them hardly fall at all for the first few hours. Mostly, the course is seen to be composite in form, with a general tendency towards being concave below, but some start with a long rectilinear slope and may maintain this as long as observed. In this type II the distinction into two subtypes is very marked.

We have undertaken a very detailed analysis of these different forms of air-line drift exhibited by individual apples of one picking, because it bears on the important question of the relative value of investigation of individual apples, as contrasted with the investigation of large representative samples of an apple population on statistical lines. It might have been that no significance could be attributed to these divergences of form shown by individual apples, and that they could only be classed as chance variations, due to causes which were too small and too numerous to be elucidated. Should this prove to be so, investigation of individual apples would be superfluous, and, indeed, tiresome. We consider that we have established the contrary position, and hope to show that practically all the features of these air-lines have a definite metabolic basis, and that the whole set of phenomena can be brought into one general system.

Our first business is to find out some other significant feature of apple respiration with which these types of drift can be correlated, and we will now take up their relation to the chronological sequence of initial intensities of respiration set out in the first section. Let us start with type III, as it is the most homogeneous. The apples included in this are XIX, XX, XXI, XXIII and XXV. Clearly they are late on in the chronology; but the omitted serial numbers XXII and XXIV, also very late, gave quite different records, so chronology is not everything. Reference to the schema of fig. 3 will show that all the type III apples come on the descending slopes of the groups A and B, and also that there are no other apples on these slopes. All these apples were yellow-green or full yellow, and would be described as nearly ripe or fully ripe. We meet here a perfect correlation—as far as it goes—between type of air-line drift and position of individual apples on our A, B, C class schema. The initial respirations of this type III vary from medium to low, but apples with these same initial magnitudes on the ascending limbs of the schema never give this type of air-line. The air-lines of XIX, XX and XXI of class A have almost identical curves, but the air-line of XXIII falls faster at first than any of them. Also the air-line of XXV is falling slightly faster than that of XXI and diverging from it in the figure. It is therefore possible to suggest that the two on the descending limb of B make a subtype III*b*, just distinguishable from III*a*, of the class A examples.

In type I we have four records, VII, VIII, XXII and XXIV, which do not fall at all over several days, but rise; so they make a very natural class. Of apple VIII we cannot say very much as it developed a fungal attack at one spot and its respiration subsequently rose rapidly, with development of visible mycelium; the initial piece alone is therefore brought into fig. 4. Apple VII gives a well-characterized record followed for 320 hr. and rising for the first 200 hr. The fall that ultimately sets in is quite unlike the fall of type III, as it starts gradually and is of increasing steepness giving a form which is concave below. Two of the apples in this rising group are early apples VII and VIII, while the other two, XXII and XXIV, are chronologically very late, so that we get no help from this consideration. The real clue is that this type I is exactly co-extensive with the class C of fig. 3 drawn up for those apples which represent a strain that ripens very slowly. Within this type we are able to make a distinction based on chronology, in that the late apples cease their rise at about 100 hr. (subtype I*b*), while VII keeps rising for 200 hr. It will be noted that all four of these apples are on an ascending slope of initials in fig. 3, and we have no knowledge of what would have happened could the research have been continued till this class of apple was fully ripe and the initials became less.

We have now to consider the more complex group of type II, which on the whole may be said to give falling curves, though not of the regular form of type III. It is clear, by exclusion, that all these apples must lie on the ascending slopes of classes A and B of fig. 3. The four earliest examples, V, VI, IX and X, exhibit the same type of form, that of a short initial stretch, which is practically level, passing into a falling stretch getting steeper and steeper. Air-lines V and VI, followed for 200 hr., show a compound form in which the form of the early part is presently repeated at a lower level, and there is some evidence that this is about to happen in IX and X, but the record was cut short too soon for proof. Then came two contemporary apples, XI and XII, of which XII conforms to the type of IX and X. The form of XI is not well established, since it was in 5% oxygen before nitrogen from hour 24 to hour 110, as shown by the long broken line representing the interpolated part of the air-line. The form suggested is a long rectilinear fall which is really the form characteristic of the next subgroup. In the apples of subgroup II*b*, which are all later examples chronologically, we find the early level initial does not appear again, but the course may be characterized as practically a long rectilinear slope from the beginning. Leaving aside XVIII for the moment, it may be noted that the distance of fall in 100 hr. is greater in this subclass than in II a. The falling tendency is thus more marked. The ends of these rectilinear falls for XIII, XIV, XV and XVI are rather obscure. The two former may be held to turn down more steeply but not so with the two others. Also, the general course is not strictly rectilinear but somewhat curved, though it lacks the very regular falling curve form of class III.

The air-line of XVIII offers an arresting contrast with that of XVII which was carried on simultaneously. While XVII gives the type of form just described, XVIII starts with a long rectilinear course sloping down but very little, so that in 100 hr. it is no further below its initial value than is characteristic of subtype II a. Later it turns over into a steepening slope. Referring to fig. 3, we see that the initial value of XVIII is much below that of XVII, and that it is therefore one of the apples that has been segregated as a member of class B. The distinction of initial values between these two apples is thus fortified by the marked distinction of the forms of their air drift. The form of XVIII has several affinities with the type of II a, which is chronologically earlier, as is appropriate.

We have now worked through all the forms of air-line drift in fig. 4 and see that they can be schematized into a system which finds the basis of its rationalization not in chronology alone or in assignment of groups A, B and C alone but in a combination of these considerations. What counts is, of course, not chronology directly but its physiological aspect—grade of senescence—and as class C is certainly ripening very much more slowly than A, and class B may be ripening somewhat more slowly than A, then the physiologically, comparable stages of senescence are displaced relatively for the three groups. The index of senescence then becomes the position of the initial respiration on the rising and falling slopes of the class lines of fig. 3. We may then, in the succeeding paragraph, achieve a synthesis of the relation between air-line drift and degree of senescence.

The least senescent apples of our population would be the earliest ones on the up-slope of the slowest ripening class, C. For this position VII is the standard and exhibits a long-continued rise, rounding off to a level preliminary to what we may style a steepening fall. The next in order should be the earliest on the up-slope of the quicker ripening class, A; and here we get no initial rise, but do get initial level courses, which in examples V and VI pass slowly into steepening falls, while IX and X go through this drift more quickly. The next phase that we expect after the initiation of this fall is a steepish rectilinear fall, and though there is some confusion of detail in this region we take the subtype II b as representing this. Apple XVIII gives support to our view, in that being class B it should not be so senescent as the class A apples of the same date, and it is quite clear that its form diverges in this direction, beginning with very little slope and passing over later into the steepening fall. The form characteristic of the next phase of senescence is very clearly indicated by all the examples of type III, where the initial steepish fall steadily flattens out in a very regular course giving curves which more and more approach a straight line. This characterizes the apples of the advanced stage of senescence, termed ripeness, and this is associated with their position on the declining slopes of classes A and B. Some support is given to the segregation of B from A, in that for a given initial value of respiration an apple on B should be less senescent than one on A, and therefore start its air-line with a steeper fall, which is a form a little further back on the general scheme. The form of the final phase of the air-line drift is revealed only by apple XXI, which was followed for a very long time. Here at the end of its record we get no further slackening of the fall, but a straight-line fall of constant slope. The fall of this slope is only $1 \cdot 2$ mg. CO_2 in 100 hr., such a slope that if it were continued at this rate the respiration would reach zero only after a period of 45 days at $22°$ C.

As we have no very senescent apples of class C, we cannot say whether they would conform to this schema, which fits classes A and B.

The regular succession of air-line forms during senescent drift suffices to establish that the effects are not the chance expressions of a multitude of small indeterminable causes, but must be the outcome of simple metabolic principles. In the next section we shall propose a definite schema of interpretation of these forms.

3. The Senescent Phase of Ontogeny and the Lowering of the Organization Resistance of the Tissues

In this laboratory various workers have studied the course of respiration in isolated plant organs of different types, and we have formulated certain fundamental principles that are to be found in action. These will be the subject of a general exposition in later papers. Here we limit our attention to a special phenomenon that we believe to be characteristic of the respiration of that late stage of ontogeny for which we have proposed the specific name of 'the senescent phase'. The special phenomenon that appears at this stage may be entitled a lowering of 'the organization resistance' of the tissues. We have coined the term 'organization resistance' to express an important aspect of protoplasmic control of metabolic rate. It is quite clear that the catabolic activity of a tissue is not merely conditioned by the *amount* of reserve food material that is present; there are times when catabolic flux is very active and times when this conversion of potential or reserve metabolites is extremely slow. This is a matter of protoplasmic organization, and it must be concluded that some of this organization is of the nature of a resistance to reaction rate. We may picture some of this hindrance to reaction as achieved by spatial separation of the reactants by impermeable protoplasmic membranes. More significant, however, will be the adsorption or combination of one or both of the reactants by the stabilized components of the protoplasm. Phenomena of this sort are presumably associated with the control of the hydrolysis of carbohydrate reserves of the polysaccharide, disaccharide and glucoside type.

A lowering of the normal grade of organization resistance would then, by definition, result in a quickening of the rate of some aspects of metabolism, more especially and significantly of those primary hydrolytic changes which bring the complex reserve and semi-reserve substances into the flux of catabolism. The particular final result of such acceleration of initial activity which interests us at present is the increase of rate of production of the effective substrate of respiration. Under prevailing conditions, in which this substrate is not already in excess, an increased production rate will reveal itself to us by an increased rate of respiration.

In place of the generalized expression—lowering of the grade of organization resistance—it will be preferable to use a narrower expression in this discussion, as it will be limited to respiration phenomena. Lowering of 'hydrolysis resistance' will serve our purpose or more conveniently the inverse of this which we may call increase of 'hydrolysis facility'. This change takes place automatically in that late stage of the life history of tissues which we label the senescent phase. We picture its onset as at first gradual and then progressing at an accelerating rate; later the acceleration diminishes, and finally the grade of hydrolysis facility ceases to increase and remains maximal at its new high level.

This senescent increase of facility takes place in stored apples at any temperature, but seems to have a high temperature coefficient so that it runs its course very much quicker at 22 than at 2° C., though the initial low and final high level of facility may be of identical pitch at both temperatures.

This conception of a fall in organization resistance and a consequent increase in catabolic changes had its origin in the search for an interpretation of the fact which we have clearly established that when the falling respiration of isolated starved organs is continuously followed, it is found that a time comes when the respiration starts spontaneously to rise again fairly rapidly in spite of continued starvation. This phenomenon of senescence will be dealt with in a more general way when we come to set out our observations on the starvation respiration of cherry laurel leaves. Here we are only concerned with its effects upon the course of the air-line drift of senescent apples.

We shall now attempt to interpret formally by a graphic schema our observed series of changing types of air-line as being the resultant expression of combined factors of senescence and starvation; the previous rate of senescence at 2° C. and that prevailing in the respiration chamber at 22° C. having both to be taken into account. In the upper part of fig. 5 are three forms of curve, Y, Z and S, representing the three significant factors. The horizontal direction of the schema represents time and the letters $b...l$ are points along the time drift that we shall be concerned with. The vertical direction represented by the distance between lines U and V represents the range of change of 'organization resistance' or 'hydrolysis facility', the level U standing for low facility and the level V for high. Before the point of time b the resistance is normal, the hydrolysis facility is therefore low, so low that the substrate of respiration would be produced at a rate to give, say, 10 mg. CO_2 per 300 g.hr. at 22° C., or equivalently, with $Q_{20°}=8\cdot0$, $1\cdot25$ mg. CO_2 at 2° C. Let us suppose that in storage at 2° C. the senescent change sets in at time b and the hydrolysis facility begins to rise slowly, reaching the maximum value of the level V at time j. The progress of this change with time is represented by the sigmoid curve Y, and at its end at j the facility level of V is supposed to be just such as to give double the hydrolysis values stated for level U. At j the senescent change is over and the facility remains at level V on to time l and beyond.

Now let us suppose that the apple at time *b* should be brought from 2 to 22° C. as its senescent phase is beginning; it is pictured that it would senesce rapidly and rise to the level *V* soon after time *d*, following the sigmoid curve *Z*. Thereafter the level would remain at *V*. Suppose in contrast that the change from 2 to 22° C. is now carried out later, say, at time *e*. By this time the senescent change at 2° C.

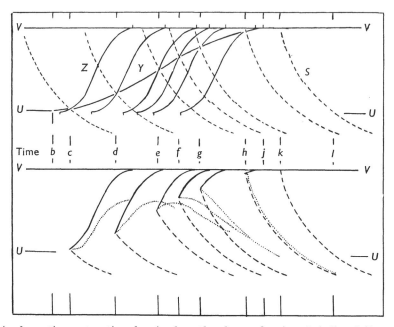

Fig. 5. A schematic construction showing how the observed series of air-line drifts may arise as resultants of opposed tendencies. For detailed explanation see text. The letters *b...l* represent a series of points of time during the progress of senescence. The level *V* indicates a high grade of hydrolysis facility and *U* a low grade. The upper part of the figure sets out as curves *S, Z, Y* the component factors affecting the drift of respiration at any moment, while the lower part gives the resultant air-line drifts as dotted curves. Curve *Y* indicates the form of the slow senescent drift at 2·5° C. of hydrolytic facility from level *U* at time *b* to level *V* at time *j*, curve *Z* the rapid drift from *U* to *V* at 22° C. A set of six identical *Z* curves for 22° C. are drawn, beginning at intervals along the time axis to fit a series of cases when apples are transferred from 2 to 22° C. at these intervals, one after the other. The significant part of the *Z* curve in each case is the part which lies above the *Y* curve, the apple being brought from 2 to 22° C. at the locus of the intersection of the formal curves. The set of seven identical *S* curves stand for the factor of 'starvation' fall of respiration at 22° C. They represent the falling tendency of respiration at 22° C. becoming effective at the point of time when the apple is brought from 2 to 22° C. In the lower part of the figure only the significant parts of the *S* and *Z* curves are retained, and these are set out for the six cases when the change of temperature takes place at *c, d, e, f, g* and *h*. The form of the resultant curve for each case is drawn as a dotted line. It will be noted that levels *U* and *V* are not absolute rates of CO_2 production, but stand for rates either at 2 or at 22° C. related to one another by the adopted value of $Q_{20°} = 8·0$.

will be half over by progress along *Y*, and the rise of temperature will cause the remaining half to be carried through quickly, following the upper part of one of the *Z* curves, being the particular one drawn so as to intersect *Y* at time *e*. The full hydrolysis facility will be reached at a time between *f* and *g*. In all, six *Z* curves have been drawn, one to indicate the rise at 22° C. for each of the arbitrarily selected series of cases when the apple is brought from 2 to 22° C. at the successive points of time c, d, e, f, g and h. In each the start of the rise of hydrolysis is along Y and the finish follows a longer or shorter track of Z.

Were there no qualifications to be made, the implication of all this system of changing organization resistance would be that the respiration rate would, if kept at 2° C. continuously, rise slowly along Y, from 1·25 mg. CO_2 at b to 2·50 mg. CO_2 at j; or, if for contrast kept throughout at 22° C., from 10 mg. at b to 20 mg. just beyond d. If, however, the apple remained at 2° C. only till time e when its respiration would be 1·87 mg. CO_2, and were then suddenly brought to 22° C. its respiration would at once change to $1·87 \times 8·0 = 15$ mg. CO_2 and then advance rapidly to 20 mg. CO_2 between f and g.

There is, however, a qualification of fundamental importance to be made which depends on the fact that, while the low rates of respiration that occur at 2° C. can be maintained, the high ones proper to 22° C. cannot be maintained, but tend to fall off by what may be termed 'starvation'. To get a pure measure of the falling starvation factor of respiration at 22° C. it is necessary to experiment outside the senescent region, at some time after j, such as k. An apple kept at 2° C. till time k will show respiration of 2·5 mg. CO_2, and if then brought to 22° C. its respiration will change quickly to 20 mg. CO_2, but cannot remain at this high level, but must fall, following the course of the broken line S, first falling fast and then slower and slower with time. The course of S represents then the pure starvation relation.

Returning to the cases where change from 2 to 22° C. occurs during the senescent phase at e or some other of the six represented points of time, then the starvation factor has to be introduced similarly at each. For this purpose identical S curves have been drawn passing through each of the points of intersection of the Y curve with the Z curves. Our components are now set out, and at any time point where there is change from 2 to 22° C. then the falling curve S at that point represents the starvation factor tending to lower respiration, while the rising curve Z represents the accelerating tendency due to increasing hydrolysis facility.

In the lower part of fig. 5 we show how these two factors interact to determine a drifting series of air-line forms which correspond in type with the series actually observed. In this lower part of the figure those parts of the lines which we may call construction lines have been omitted, and at each of the six points of time we have left in the figure only the two opposed factors brought into being, but into opposition, by the change to 22° C. The form of the resultant curve arising from this opposition has been constructed for each of the six points by calculating the difference of the upward rise and the downward fall for a succession of short lengths of time. These forms of air-line drift are set out one by one as dotted lines.

In the early ones, while the facility is rising fast the resultant air-line drifts upwards at first, while if started at f it runs a level course, and if started later falls all the time. In each single case the rising facility component has a less and less effect as it nears its end at level V, and this effect becomes zero when the change is over; after this point of time the air-line follows a pure starvation course, being determined by the appropriate region of the S curve alone. The seventh curve of the series, which begins at k, has by definition no rising component, and is the pure S curve throughout its course. The series of air-line drifts that we have synthesized in this schema presents all the observed types that we have set out in fig. 4, p. 12. The drift of k is the representative of type III (see table, p. 12), but this is of course due to our definite selection of this form of starvation curve for our schema. The resultant curve d is the analogue of type Ia, in which the rise is long continued, and curve e the analogue of type Ib. Then as we follow on, curve f with its level start is the analogue of type IIa, while curve g giving a rectilinear fall represents type IIb.

Our work on the apples was not begun until they had been in storage 6 weeks, and we consider that the absence of any observed air-lines rising so steeply as the resultant curve c is due to the fact that this type occurs only in the earliest stages present in October.

It should be stated that this schema is not an attempt to reconstruct the actual air-lines observed but only the *types* in their proper serial succession. The curves Z, Y and S, employed in building up the schema, were not arrived at by careful trial of form to give the best fit, but were drawn freehand without subsequent adjustment. Nevertheless, it is obvious that, for a given time axis in the schema, the mutual relation of the slopes of the adopted Z and S curves determines the synthetic form of the air-line drifts. Had S been nearly as steep as a vertical line, or nearly as flat as a horizontal line, Z being unchanged, then the air-line forms would come out very different. Some narrow range of relation between steepness of S and Z has therefore really been predicated in the application of the schema. More general aspects of the significance and forms of starvation curves have to be taken up in a later paper.

Further, it will have been noticed that nothing has been said, so far, about the existence of a starvation curve component at $2°$ C. as well as at $22°$ C. We have tacitly assumed that this component at $2°$ C. is so flat in form that it involves so little decline of respiration rate with time that it has a negligible effect upon the slope of the air-line. At this temperature, then, the rising facility curve is held to express itself fully stage by stage in the observed rate of respiration. We have made no respiration measurements at such low temperatures, but this aspect of apple metabolism has been studied by Kidd and West, and the relations we have indicated may be put on a more exact basis when their work and ours come to be correlated.

In all this matter we have treated the individual apple as a whole and have not stated whether the drifts observed are to be considered as true for each individual cell of the apple or only as a statistical truth for the drift of the whole population of cells making up the tissue.

At least it can be said that our schema fits the facts, and that it provides a new conception which helps us to interpret the complex behaviour of respiration phenomena, by assuming that the organization resistance of tissues is not constant but is capable of undergoing spontaneous change. In other work we shall show that the resistance can be altered by experimental treatment.

4. The Initial Effect of Change of Temperature

When an apple is heated up from 2 to 22° C. the main effect upon the respiration is an increase of rate to somewhere about eight-fold. The course of this rising respiration, hour by hour, does not, however, proceed to the new rate by a continuously rising curve of the form that a simple transition from a low steady rate to a higher steady rate would give, but the transition record may exhibit a definite peak, so that the CO_2 production is, for a time, in excess of the air-line rate that it will attain later on. In the present section we have to examine the early parts of our records rather carefully for evidence of the presence and magnitude of this effect.

For this purpose these parts of the records have all been brought together in the two columns of fig. 6, alined to zero hour; and certain construction lines have been drawn upon them to facilitate their examination. For clearness, in this figure, each record is presented by its pair of 'contour lines' (see p. 4), and the actual readings lying between them are omitted. In the earliest steeply rising hours the contour lines lie close together, but later they are at the standard distance apart due to the fluctuation of the respiration (see p. 4). In this later region a median line is drawn between the contour lines giving the direction of drift of the specific air-line of that individual apple. We have extrapolated each of these air-lines back to zero hour to provide one of our two construction lines. At the top of the figure, two in each column, will be seen the four apple records which exhibit a definitely rising air-line, and in three of these, VII, VIII and XXIV, the existence of temporary excess values of respiration is very clear. After the initial maximum the record shows a series of declining values, which present the curious feature of lying on a straight line, and this presently brings the record on to the true air-line. Here then there is a well-marked 'inflexion point' on the record as the respiration values thence proceed to rise along the air-line. As a second construction line in the figure, the median line of the observed falling slope has been continued back to zero hour, and onwards in the other direction to define its slope more clearly. We thus get superposed on the record a pair of construction lines intersecting at the inflexion point. The angles at intersection appear to be constant in all the well-marked cases, being 25° and 155° with the relation of axes adopted in this graphic presentation. The slope of the passing-off of the initial temperature effect is not then a constant slope to the horizontal, but a constant difference of slope above the slope of the air-line of the individual apple. One implication of this relation is, of course, that the higher the initial maximum lies above the extrapolated air-line, the later it will be before the inflexion point is reached. The time values for these relations are set out in the table in Appendix I.

It now becomes interesting to consider our numerous records in which the air-lines show a well-marked fall as a general character (see § 2). These eight records are grouped in the lower part of fig. 6, four in each column. Here we can hardly expect to find so obvious a contrast of direction at the inflexion point, since both lines will be falling ones. The method we have adopted for interrogating each record is that of superposing on it tracing paper on which our two lines,

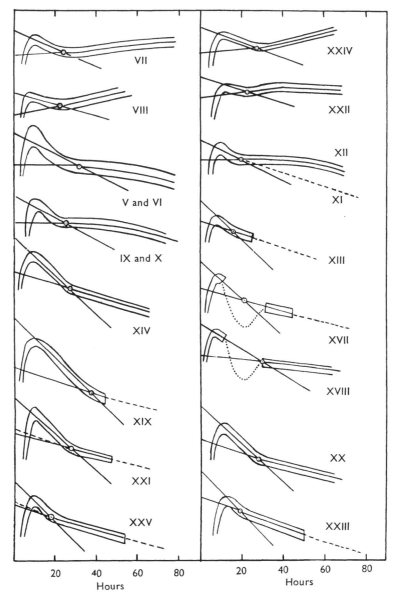

Fig. 6. The initial hours of all the records brought together for examination of the 'initial temperature effect' on change from 2·5 to 22° C. The low readings of the first 6 hr. are omitted to save space. The actual course of respiration is defined by the track of the double contour lines. The ordinate scale and the values of the single readings can be seen in the charts of the records in Appendix I. The circle on each record marks the 'inflexion point' and indicates the end of the initial effect. The median air-line is extrapolated to zero hour. Where this air-line has a marked curvature the extrapolated curve is given as a broken line, and the continuous straight line near it is a construction line drawn as a tangent to the curve at the inflexion point. A second construction line is the straight line drawn through the inflexion point, along the median track of the record before the inflexion point. For the application of these construction lines see the text. Records XVII and XVIII are interrupted by a failure of temperature control from hours 12 to 32 (see Appendix I).

intersecting at the standard angle, have been drawn. One line is put to coincide with the course of the air-line and shifted along it to see if the intersecting line comes to lie on the median course of the falling slope of the change of temperature effect. It will be seen that apples XIV, XX and XXI give very convincing evidence of the existence of a big effect when ruled up in this way, and the intersection point is correspondingly late, 27–29 hr. (see the time values in the table in Appendix I).

Two other records, XXIII and XXV, show small effects, about which one cannot be very certain, the inflexion point coming at 18–19 hr. Apple XIX appears to have an exceptionally large effect with inflexion point at 38 hr., but this apple was transferred to nitrogen at hour 44, so the air-line course after the inflexion point is not very securely located; on general evidence we believe the dotted track in the figure is the course it would have followed had it been kept in air. Apples XVII and XVIII, investigated together, both show an unfortunate break in their records, due to an accidental fall of temperature in the bath soon after the point of time when the maximum of the effect had developed at hour 9. The effect of this lowering of the respiration was not fully recovered from till hour 32, and a single median track only is drawn for the respiration of this distorted period. The direction of the subsequent air-lines for these two experiments enables us to extrapolate to zero with confidence, and our cross-lines can be fitted to the two fragments in a way that is satisfactory enough. We may conclude that with all these cases of respiration the same type of change of temperature effect occurs as with rising air-lines.

We have still left a group of four records in which the air-line starts with a short level course. These, which appear in the middle of fig. 6, show a well-marked temperature effect, and the same construction of the intersecting lines has been superposed upon the records. Though there is no doubt of the presence of the effect in records V and VI, IX and X, yet the fit of the straight lines is not so convincing as in the other cases. Both these two records are composite records, that is to say, that all the readings for two separate experiments V and VI or IX and X, carried out simultaneously, have been plotted as one record. In each case the readings of the pair intermingle, and their collective fluctuations fall within contour lines which are not twice as far apart as the normal spacing for a single apple (see the records in the charts of Appendix I). It will be noted that the falling slope from the maximum to the inflexion point in these four apples can hardly be described as a straight-line course, but tends to be a declining curve, which indeed is the form to be expected *a priori* in all cases were the fall to the air-line to be of an exponential nature. Apples XI and XII also start with a short-level air-line, but the temperature effect is not very large and so is not very well defined, but here also there is some indication of a slope from the maximum which is convex below rather than rectilinear. We would conclude that in the group of apples which start with a level air-line the same type of temperature effect takes place as in the other two groups, but that here we have a variation of form of the slope from the maximum. It interests us, in our endeavour to analyse out all the details of behaviour of apple respiration, that this group of apples is not a random one but is the group of apples with air-lines of type II*a* on p. 12.

The only records that we have not yet referred to are XIII, which is so short that it is ill-defined but appears to exhibit a minimal special effect, and the pair XV and XVI for which, through an accident, we have no early readings at all. The change of temperature effect, then, may be held to be always present in our apples, but its magnitude varies a good deal. Its uniformity of form is striking, seeing that it keeps true to type, whether respiration, fundamentally, is either rising or falling, while the departure from the majority type is but slight with level respiration. Possible interpretations of this effect will be examined in the next section.

5. Change of Temperature and CO_2 Production

Let us first consider the various factors within an apple that might express themselves by a transient excess evolution of CO_2 to the external air current when the temperature is quickly raised from 2 to $22°$ C. The effects of this heating up upon CO_2 evolution may be divided into physical and metabolic; and two processes may be considered under each heading.

Physical processes: alteration of equilibrium of solution, adsorption and loose chemical union of CO_2. At the initial temperature of $2°$ C. the apples start with a certain percentage of CO_2 in their air spaces plus so much in solution in water in equilibrium with this, so much CO_2 adsorbed and so much loosely united as bi-carbonates, etc. All these states are reversible ones, and the equilibrium point of each will be shifted to a lower value in the sorbed phase by a rise of $20°$ C. in the temperature. From each state then CO_2 will tend to be liberated by a rise of temperature, and it might be thought that the excess production of CO_2 that we have noted was simply an outcome of the shift of these equilibria. Thus the absorption coefficient of water, when heated from 2 to $22°$ C., falls from $1·6$ to $0·8$; so that the water would give off half its dissolved CO_2 provided it continued in contact with the same external partial pressure of CO_2. On further consideration of the conditions established in the heated-up apple, it appears, however, that this provision is not complied with. For an apple at $22°$ C. gives off steadily by respiration to the air current some $8·0$ times as much CO_2 as the apple at $2°$ C., and as the ultimate stage in this escape is inevitably a diffusion gradient across the surface of the apple there must be maintained a much higher concentration of CO_2 inside this surface at 22 than at $2°$ C. for such a transference by diffusion. The increase of diffusivity and decrease of viscosity in the medium are not very great for this rise of $20°$ C.

In accordance with this expectation it has been found by work to be published subsequently that the internal atmosphere of an apple contains at least five times as much CO_2 at the higher temperature. The rise of internal partial pressure of CO_2 is thus so great that it actually overbalances the shifting of the solution equilibrium point. It follows that the water in a respiring apple at $22°$ C. should contain in solution at least $0·8/1·6 \times 5 = 2·5$ times as much CO_2 as at $2°$ C. We cannot, then, attribute any of the evolved excess to this source.

Another possible physical source of CO_2 would be the liberation of adsorbed CO_2 by the rise of temperature. The same general considerations must be borne in mind as for the case of solution in water. The actuality of liberation will depend upon whether the known rise of partial pressure of CO_2 within the apple to about five-fold is adequate to balance out the lowered affinity of the adsorbent. There is no quantitative knowledge available for the apple; also its content of organic matter is low. It is to processes of a metabolic nature that we incline, therefore, to attribute the observed change of temperature effect. To the two metabolic processes that suggest themselves we may give the names of 'carbohydrate equilibrium effect' and 'intermediate compound effect'.

The carbohydrate equilibrium effect. It is well established that such a change of temperature as our apples are subjected to produces a shift in what is generally called the starch-sugar equilibrium relation, so that sugars accumulate at the expense of starch during the time the tissues are kept at low temperatures. When the tissues are brought to a higher temperature this accumulation of sugar is reconverted to starch so that initially the sugar concentration is in excess for the high-temperature state but rapidly declines. We should expect, as a result of this sugar behaviour, high initial respiration at 22° C., falling with time as the sugar falls. This respiration effect is very marked with potatoes on change of temperature and the apple effect might be of this nature. There is also evidence that such temperature changes affect the relations between cane sugar and hexoses. The respiratory effect in apples might be due to this metabolic cause.

The intermediate compound effect. Picturing respiration as fundamentally a sequence of linked reactions its progress at a steady state must be associated with definite equilibrated concentrations of the series of intermediate compounds that constitute the reactants. With rise of temperature and the change from a steady state at 2° C. to a new steady state at 22° C. there are possibilities of transient excesses of CO_2 production according to the varying effect of the rise of temperature upon the component reaction velocities.

We have before us, then, three possible mechanisms that may contribute to the excess production: the adsorptive, which is outside the cell's metabolism; the carbohydrate equilibrium, which is metabolic but outside the essential respiratory nexus; and the intermediate respiratory compound, which is of the essence of the respiration itself. It would seem that these three mechanisms have sufficient *differentia* for us to hope, later, to distinguish between them. All we can do at present, as a contribution towards solution of this problem, is to survey the set of effects that we have recorded and determine whether the variations of magnitude and timing that they present give any helpful indications. Some indication might be obtained should a clear correlation appear between the varying size of these effects and some other feature of the apple respiration. Were the effects of identical magnitude in all cases, then this relation might be held to support the physical interpretation. Should there be a close correlation between intensity of individual respiration and the magnitude of the effect, then the intermediate respiratory compound view would find support. If, on the contrary, the effects were large at early stages of senescence and steadily declined chronologically so that there was a correlation with what we may call 'starvation', then this would support the carbohydrate equilibrium interpretation. We shall see presently that no single one of these possible correlations dominates the whole situation.

One considerable difficulty that stands between the observer and elucidation of these transitional phenomena is that the intensity and duration of an intracellular production of CO_2 is so much distorted when the observed signs of it are only the intensity and duration of outward escape of CO_2 by diffusion through the surface of the massive tissues of an apple. Should a few cubic centimetres of CO_2 be suddenly produced in excess by the respiratory mechanism, which before and after maintained a steady state, then this would not manifest itself outside as a sudden output of CO_2 of identical timing and intensity, but as an output which might rise fairly quickly to a maximum but would decline quite slowly again towards the steady rate, owing to diffusive lag. The decline would take the form of a 'logarithmic curve'. It is clear then that when we observe the CO_2 production of an apple falling in a curve of this sort from a high level to a lower level, and taking many hours to complete the falling transition, we cannot conclude that any actual internal production of CO_2 has been continuing at a heightened rate beyond the point of time at which the high level ceased and the fall began. The observed escape of CO_2 may be described as a distorted anamorph of the production.

A measure of the total magnitude of the excess internal production of CO_2 is, of course, given by the total excess escape of CO_2, provided our graphic records provide a sufficiently certain base-line above which to measure the area expressing the excess escape. This certainty we do not possess in dealing with the initial temperature effect in the apples, but we may take a summary survey of the magnitudes in excess of the base-line which is provided by the theoretical extra-polated course of the air-line. This survey is represented in fig. 7, where the common base-line stands for the air-line level and each magnitude is represented as the area of a triangle emerging above this base. The vertical side of each triangle gives the height above the air-line value, which the CO_2 output reaches at its maximum, and the length of the triangle along the base-line measures the time between the maximum of the effect and its extinction at the inflexion point on the air-line.

In this figure are assembled a series of these triangles, and we have to inquire whether their respective sizes fall into any simple system. After a careful survey of the cases it seems to us that the distribution of magnitudes has a dual determination, and that it is first of all essential to take account once more of the A–B–C class schema evolved in the first section. In fig. 7 the individual cases have been grouped into three rows for the three classes and spaced along the rows according to their positions on the rising and falling slopes of their respective classes. In class A we find apples V and VI, IX and X and XI and XII, which are on the rising slope, presenting a clear chronological drift of declining magnitude of the special effect, so that here there is an inverse relation to rising magnitude of general respiration and a direct relation to progressive starvation. Further along on the falling slope of class A we have apples XIX, XX and XXI, and these show another declining series with time and progressive starvation, but here, in contrast, the general respiration pitch is falling with time. Between these two groups the special effect must have increased in magnitude, but the only available apple between is XVII, so there is no clear evidence of the real course of the facts in this region. In class B we have a suggestion of a parallel to the second falling group in the relation of the series XVIII to XXIII and XXV. Apple XIII provides a very low value in the middle region. Class C, with its long-continued rising slope, provides a diminishing series of effects from VII and VIII to XXII followed by the large effect of XXIV.

Viewed in this way it may be held that the whole set of effects shows evidence of decreasing magnitude with progressive starvation up to the maximum of initial respiration, but that here the end of the senescent rise of general respiration brings

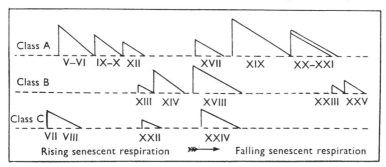

Fig. 7. A diagrammatic presentation of the relative magnitudes of the excess CO_2 production in the 'change of temperature effect' for the nineteen recorded cases. The areas of the triangles provide the relative measures of magnitude, the vertical axis being intensity and the horizontal duration. The base of each triangle measures the time of the falling part of the effect from the observed maximum to the end at the inflexion point. The vertical of each triangle measures the intensity of the excess production above the arbitrarily adopted values of the ideal air-line; the true base-line from which the excess should be measured cannot at present be determined. The triangles are arranged in three series, corresponding to the three classes of rate of ripening, and are spaced out within the classes so as to bring corresponding states of senescence over one another. Thus arrayed we find indication of an initial declining series followed by a reversion to high magnitudes and then a second declining series.

about metabolic change to a new basis, and the magnitudes rise only to decline once more as starvation progresses further. For this interpretation of the magnitudes of the initial temperature effects as a double series of drifts we do not claim even high probability, but it provides suggestions for future examination. At least it seems clear that the observations cannot be rationalized without recognition of class grouping and of the main phases of senescent drift.

There are a good many factors that might be significant in the interpretation of these effects. We have had the advantage of additional suggestions from our colleague Mr Briggs in discussing this and other biophysical aspects of the present stage of our general analysis of respiratory phenomena.

CONCLUSION

In this paper there has been taken up the task of treating analytically the respiratory phenomena presented by a collection of apples in storage which are slowly ripening during a period of eight months. The general metabolic phenomenon that is proceeding in such a population is a steady drift through a definite onto-genetic phase, which, to distinguish it from the phases of adolescence and maturity, we term the senescent phase. This phase consists, in essence, of a fundamental change in the organization of the tissues, which we describe as a lowering of the normal organization resistance, so that hydrolysis of reserve and semi-reserve substances proceeds at a faster rate than in the mature phase. This change leads to a greater production of the effective substrate for respiration, and so to an increased production of CO_2. When this senescent change has completed itself respiration falls in the direction of zero by the natural starvation condition that is present in an isolated plant organ.

Our procedure for the analytic study of this stored population kept at about $2 \cdot 5°$ C. has been to remove from it individual apples at intervals and to examine their respiration at $22°$ C. The primary quantitative features that the respiration of each individual presents are two: (1) the initial value which is the measure of its physiological state when removed from store, and (2) the form of the subsequent course of respiration hour after hour when continued at $22°$ C. These two features exhibited a great variety of magnitudes and forms, and at first sight the distribution of these in chronological series, apple after apple, seemed to be almost random. After careful critical collation of all the forms, evidence of systematic drift emerged.

Finally, the conclusion was forced upon us that there were certainly two, and possibly three, physiological classes of apples present, which classes ripened at different rates so that chronology and metabolic drift of the whole population did not move together. On correcting for this we could formulate the characteristic phases of the metabolic senescent drift, and show their succession in each class. We found then that the observed respiration of an apple is an integration of two independent and opposed processes that are at work in senescence. One is the starvation drift at $22°$ C., tending continually to lower the respiration, while the other, tending to accelerate the respiration, is the lowering of organization resistance, expressed in this connexion as rise of hydrolysis facility.

In the third section of the paper a scheme has been presented in which these tendencies are brought to account, and it is shown that the observed types of phenomena can be reconstructed in this way. We would not claim that the inter-pretations that we have put forward in this paper are fully established by the evidence produced. We set ourselves the task of examining every detail of respira-tory behaviour, and after collation of the details drawing up a schematic inter-pretation. As the apples were steadily drifting into new metabolic states, one after another, all the time, it was not possible to go back and recapture any precise set of conditions. Experimental verification of a hypothesis was therefore not a practicable procedure. Only by working over the whole field again in another year will it be determined how far the details of interpretation can be maintained.

II

THE RESPIRATION OF APPLES IN NITROGEN AND ITS RELATION TO RESPIRATION IN AIR*

CONTENTS

INTRODUCTION

The previous paper of this series contains a study of the respiration of a set of twenty-one apples brought from cold storage at 2·5° C. and examined in the laboratory at 22° C. The complex phenomena presented by the respiration while in a current of ordinary air were alone dealt with in that paper, but these same individual apples were also investigated in nitrogen and in various percentages of oxygen. In the present paper the respiration data obtained in nitrogen are to be examined, while subsequent papers will take up the phenomena presented in concentrations of oxygen.

The complexities of CO_2 production in nitrogen are no less than those already studied in air, though they bring us into a new field in which the oxidation factor is excluded. The behaviour of a number of individual apples alternately in air and nitrogen is recorded by a long series of continuous observations, and these call for detailed analysis.

So many relations which seem both novel and significant emerge from this analysis that we propose to set out our study of them formally in three successive parts. Part I deals with the actual course of CO_2 production in nitrogen and presents our observations empirically as a set of 'nitrogen effects'. In Part II our data are presented in the way usually adopted by workers in this field, in that the magnitudes of CO_2 production in nitrogen are set out as ratios to the CO_2

* This paper was published in 1928 as the second paper in the series 'Analytic Studies in Plant Respiration'. The author was P. Parija, the reference being *Proc. Roy. Soc.* B, **103**, 446, 1928.

production in air. This stage of analysis points the way to a more fundamental examination, which is carried out in Part III. In that part an attempt will be made to interpret the observed phenomena as manifestations of the working of a catalytic system of respiratory reactions, bringing into account both the main aspects of respiration, namely, those of oxidation and those of carbohydrate consumption. This part is presented as the third paper of this series.

Experimental procedure. The general conditions of experimentation are fully described in the first paper and need not be repeated. For the nitrogen experiments the procedure adopted was to disconnect the air stream through the apple chamber and to substitute a stream from a cylinder of compressed nitrogen, the flow being adjusted by a valve to give the standard rate of 1500 c.c. per hr. The stream of gas issuing from the chambers passed to a set of Pettenkofer tubes in which the CO_2 was absorbed by baryta; the production of CO_2 is expressed as mg. per 3 hr. per 100 g. fresh weight of apple.

It was found that some cylinders of commercial nitrogen contain as much as 1 % oxygen. As very small partial pressures of oxygen give quite different results from oxygen-free nitrogen, the gas had to be freed from traces of oxygen by slow passage over heated copper gauze on its way to the chamber. This gauze was contained in a silica tube in an electric furnace heated continuously to 700° C. After the furnace came a Pettenkofer tube of baryta solution to remove any traces of CO_2 before the gas passed to the apple chamber.

Part I. An Empirical Survey of the Effects of Nitrogen upon CO_2 Production in Apples

The records. Among the full records of the respiration of individual apples set out in Appendix I will be found ten cases in which individual apples have been subjected to nitrogen for periods varying from 40 to 100 hr. and their CO_2 production compared with that in air. Nine of these cases bear, in chronological order, the numbers VII, XI, XIX, XXI*a*, XXI*b*, XXII, XXIII, XXIV and XXV*. In this list appear no more than two out of the first eighteen apples, while five out of the last six were examined in nitrogen. Behind this unequal distribution lies the story of a quite unexpected complication. The record of the first apple treated with nitrogen, no. VII, was evidently of quite the same type as we were familiar with from a detailed study of the behaviour of cherry laurel leaves in nitrogen, so this section of the work was dropped for experiments on small percentages of oxygen. Not until apple XIX was a second experiment made on respiration in nitrogen, and to our surprise the type of record obtained was quite different from that of VII. This divergence concentrated our attention on nitrogen for the rest of the series, but the lack of more nitrogen records in the first part of the series leaves several problems unsolved for the present.

---

* The tenth case, that of apple VIII, was exceptional, being attacked by a fungus mycelium, and the complex problem of the effect of nitrogen on the respiration of the combined system is postponed to § 4.

The numerous nitrogen experiments made after XIX showed that there were really two well-characterized types of nitrogen effect.

Of the nine records in nitrogen to be considered, seven were cases in which the nitrogen treatment was both preceded and followed by a period in air, and these

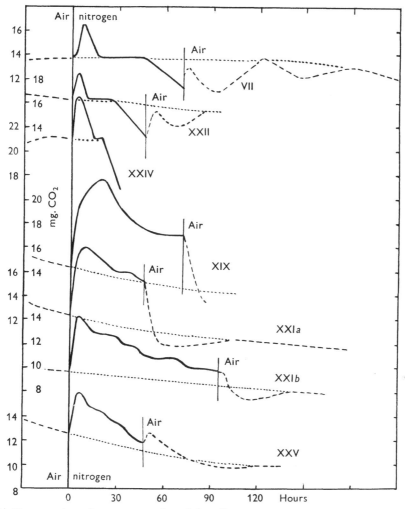

Fig. 1. Presentation of seven examples of the effect of nitrogen upon CO_2 production, being all the observed cases in which nitrogen was preceded by air. All cases alined for hour of entry into nitrogen. The heavy continuous line is the record of the CO_2 in nitrogen. The broken line before and after nitrogen is the record of CO_2 in air. The dotted line gives the 'air-line respiration' by interpolation. For the full records see the charts in Appendix I. Groups of the appropriate ordinate values are given to the left. The top three records are those of class C apples; the bottom four those of classes A and B.

must first occupy our attention, leaving the two complicated cases in which the treatment was different for § 3.

Fig. 1 presents a chart of the directly significant parts of the records of these seven apples, arranged so that the hours of beginning of nitrogen for all cases are in alinement. The chronological order of the experiments was that of the Roman numerals attached to them, but they are not arranged in that sequence but are grouped into the two different types of nitrogen effect. It will be seen that the top three in the chart differ in general appearance from the bottom four. One of the main puzzles, on the first survey of results, was to find out some explanation of why, apparently at random, one apple gave one type and another the other type. The difference could not be directly correlated with immaturity and maturity, because the first apple examined in nitrogen, VII, in November, gave the same type as two of the last examined (XXII and XXIV) in June of the next year. Those examined intermediately (XIX, XXIa and XXIb) gave the other type of reaction. Nor could the difference be correlated with differences of high and low magnitudes of respiration in air, for VII has a low air value, while XXII and XXIV have high ones. Apple XXV alone gave a reaction that could, to some slight degree, be regarded as intermediate between the two types.

The solution came, not from a study of the nitrogen records themselves but from a detailed analysis of the behaviour of the respiration of individual apples in air. This has been carried out in Paper I, where the conclusion has been arrived at that these apples were not really a homogeneous population, but that they must be segregated into representatives of certainly two, and probably three, physiologically distinct classes. Class A, distinguished by quick ripening, is sharply marked off from class C, which ripens very slowly, while there is some evidence for a third class, B, which ripens at an intermediate rate. It is a great support to this conclusion to find that this classification gives the key to the distribution of the two types of nitrogen effect.

We have now to bring into prominence the striking differences between the two types by characterizing each of them. In the records brought together in fig. 1, the abscissal hours are numbered from the change into nitrogen from air; and the continuous line gives the course of the CO_2 production that results, expressed in mg. CO_2 per 300 g.hr. of fresh weight of apple. The figures on the ordinate axis adjacent to each record give a group of the numerical values appropriate to that particular record. The broken line for 30 hr. before zero time gives the course of respiration in air during that period, while the broken line after the continuous line marks the course of respiration in air when the apple is returned to this gas. The dotted line which continues the first broken line through the period in nitrogen is the course that, we conclude, the respiration would have followed in air had not nitrogen been substituted for it. These pieces of the line together constitute the 'air-line' of the record. It will be noticed that the interpolation only joins the actual air values some way on after the apple is back in air again— not till there is evidence that the respiration is really in adjustment with air again and the transitional after-effects have passed away.

1. *The nitrogen effect with apples of class C*

To this class belong the records of apples VII, XXII and XXIV, presented at the top of fig. 1. In all three of these records it will be seen that when the apple is subjected to a current of pure nitrogen the resulting change of CO_2 production follows a highly characteristic course. It rises at once for a few hours and then falls sharply to a level identical with that of the air-line; after this level value has been maintained for a longer or shorter period of time, there sets in, quite suddenly, a rapid decline of the respiration in a falling straight-line course.

The whole observed record for apple VII lasted no less than 11 days of continuous estimations, each of 3 hr. duration (see Appendix I for the full record). Before nitrogen was given the apple had been observed in air already for 94 hr., but only 30 hr. are brought to this special figure. The air record before nitrogen is nearly level, with a faint rise in it. After the change to nitrogen there is a rapid rise, giving a crest with a maximum of 16·6 about 9 hr. after the change. This outburst of CO_2 subsides as quickly as it rose, and in 20 hr. from the beginning of nitrogen it is down to the original level that it held in air. At this level it now remains from hour 20 to hour 48, when it starts a new change and declines fast in a straight line, reaching 11·3 mg. at hour 68. At this point its nitrogen experience was terminated by a current of air.

The air record after nitrogen never rises above the 'air-line' and there is apparently a good deal of metabolic disturbance following the exhibition of nitrogen in the record of VII, for the CO_2 values follow a fluctuating course, drifting first up, then slowly down and up again, before the air-line is reached once more, in a couple of days. It is not proposed to interrupt the exposition of these nitrogen effects by justifying in detail the course given in the figure to each air-line. These are considered in the notes to Appendix I containing the whole records.

Not till apple XXII did we get another record belonging to the type of class C. The respiration in air was 16·1 when nitrogen was given, and rose to 18·2 in 5 hr., it then fell to 16·1, completing the hump in 12 hr. After this there comes a level course, lying on the air-line, lasting till hour 28, at which time the steep downward track sets in, reaching 12·7 at hour 50, when air was given again. The behaviour after air is identical with that of VII as far as it was followed. Like VII, the fluctuations start with a sharp rise, but not enough to bring the record above the air-line. The chief difference from VII is that the level second phase in nitrogen is so much briefer in XXII, 16 hr. instead of 27 hr. This apple XXII was green-yellow and nearing maturity, while VII, taken six months before, was unripe green. This suggests a decreasing resistance to anaerobic conditions with advancing senescent development.

Seeing how comparatively rare this type of nitrogen effect had been up to experiment XXII, it was surprising to find that one of the next pair of apples to which nitrogen was given, XXIV, belonged again to this type. The respiration before nitrogen had been followed for 100 hr. and had been rising for 85 hr.; then it became nearly level for 15 hr., reaching 20·8 when nitrogen was given. We now get the characteristic sharp hump of 16 hr. duration, but here rising higher, up to 24·4. This was followed by signs of a level phase on the presumed air-line, but this only lasted 6 hr. and suddenly down plunged the line of CO_2 production. In this case it was decided to determine whether this fall was in any way a toxic effect, for no apple had been killed by nitrogen among the previous experiments.

Nitrogen was therefore kept on for 112 hr. (see the full record), and the fall continued straight down to the value of 5·3 and then sloped off less steeply to 2·4. When air was readmitted the respiration only rose to 4·0, so that this apple has been practically killed by absence of oxygen for 112 hr.

With these three cases it is difficult to avoid the conclusion that as apples of this type slowly ripen, their metabolic working changes so that they become in some way decreasingly resistant to nitrogen treatment.

One more nitrogen experiment, XXV, was tried before the work had to be brought to an end, but this turned out not to belong to the type of class C.

The three apples VII, XXII and XXIV are, then, the cases on which a special type of nitrogen effect was established. It was only later that it was made out, from study of the air records and apple colour, that there were classes of apples of different ripening rates, and then it was seen that the slow-ripening class C was just co-extensive with the group that gave this cherry laurel-leaf type of nitrogen reaction, and could not resist, without depression below the air-line, long withdrawal of oxygen.

It may be mentioned here that on consideration of the early apple VIII, which became attacked with mycelium and is therefore to be dealt with separately in § 4, we came to the conclusion that it also was a representative of class C.

2. *The nitrogen effect with apples of classes A and B*

There are in all six records which conform to this type of nitrogen effect, but in this section we shall only consider XIX, XXI*a*, XXI*b* and XXV, postponing to the next section the cases of XI and XXIII, because these two were not in air immediately before exposure to nitrogen, but in other gas mixtures. Their forms, therefore, need more interpretation than those now before us. All these four cases, as their high serial numbers show, came late in the season, and even the earliest, XIX, had begun to turn yellow in cool storage. On reference to fig. 3 in Paper I, which presents a schema of the senescent respiratory drift of the three classes, it will be seen that all these individual apples, whether belonging to class A or class B, are well down the descending slopes of the schema. Had the main lines of grouping of the apples been known beforehand, care would have been taken to investigate also in nitrogen the less senescent apples on the ascending limb of the schema.

Our immediate object in this section is to compare these A and B records, determine the common elements in their form, and show how much their appearances contrast with the typical form of class C. The graphic treatment will be similar and the four records are to be found at the lower part of fig. 1.

We may first consider the records of apple XXI*a* and XXI*b*, as this apple underwent longer examination than any other case. This apple had been observed in air for 50 hr. before nitrogen, and the course of the air-line was well established, having fallen in a steepish curve to the value of 12·4 (XXI*a*). With nitrogen there is a big immediate rise of CO_2, reaching the top of a hump at the value of 18·2 in 10 hr. and then declining in a fluctuating curve, but all the time well above the air-line. At hour 48 air was given again and there is a rapid drop in the record. When, however, the value reached that of the air-line it did not stay there but carried on, giving a marked dip below the air-line values. This fall gradually slackened off and presently ceased, after which the values slowly mounted to the air-line and then continued along it. The drift of the air-line is here well established, as the apple remained in air for a further 140 hr.

Comparing the records in air, on each side of the nitrogen experience, we see that they can be joined up to give one regular continuous air-line drift, and it appears that nitrogen has had no permanent disturbing effect whatever. It was decided, therefore, to give an even longer exposure to nitrogen with this same apple, and record XXI*b*, which, in the general records, should be directly continuous after XXI*a*, is shown in fig. 1, alined under XXI*a*, for comparison. The air value of 9·7, at which nitrogen was given the second time, comes just about a week from the beginning of the XXI*a* record. Still we find exactly the same reaction to nitrogen, though the hump is here not so high absolutely. The exhibition of nitrogen lasted 97 hr. this time, but there is no trace whatever of the toxic effect of nitrogen found in apple XXIV of class C.

The long record of XXI*b* is instructive as demonstrating that the main drift in nitrogen does continue on as a downward slope which converges on the air-line. The track clearly follows an undulating course, fluctuating on either side of the ideal line that can be drawn smoothly through it. The fluctuations get less and less as the pitch of the line declines. Air was given at hour 335, and the record in air is again of the same type as at the end of XXI*a*, passing through a dip below the air-line at about 18 hr. The whole experiment had now lasted 17 days, and further exhibition of nitrogen on it could not be tried.

We may now turn to apple XIX, which preceded XXI. The respiration in air was first observed for 45 hr. and fell fairly rapidly, reaching 14·2, at which point nitrogen was given. There follows a very pronounced hump, reaching the high level of 21·6 at hour 22 and falling in an undulating track till hour 72, the respiration in nitrogen being still greatly above the air-line. At this point air was given causing a rapid fall, and the record cuts the air-line in 12 hr. and then dips below it. The temperature adjustment of the thermostat failed here and the experiment was discontinued, but the record, as far as it goes, has exactly the same features as the records of XXI*a* and XXI*b*. As the record was not followed on to return to an air-line, the exact location of that air-line cannot be established, but reviewing all the evidence its course cannot be far from that drawn in the figure

The last apple of this class to be examined is XXV, which was definitely selected from the cool store as being the ripest and yellowest apple showing no trace of brown that was present at that date (21 June). Apples of the C class were still greenish, and most of the A apples had rotted, and there is some evidence for regarding XXV as belonging to the B class, which is held to ripen a little slower than A but much faster than C. This case was observed in air for 55 hr., during which time the respiration fell gently to 12·5. Nitrogen was then given and produced the usual quick rise, to a maximum of 15·8 in 7 hr. Then the respiration declined quickly by a slightly wavy record for 48 hr., rapidly converging on the air-line. When air is given the record shows a new feature, in that the CO_2 production first rises to give a small hump and then slopes away slowly to just below the air-line, afterwards rising gradually to it. It is curious to find that the superficial form of the record in air after nitrogen, in this case, more closely resembles its own record in nitrogen after air than any other record of air after nitrogen. Clearly this apple has some special features, and at first it was thought that these might be characters of class B, but again it is the extreme end of the ripening series and the special features might be the expression of this position; the latter is the view that we adopt when in the next section we come to consider yet another apple of class B, namely, XXIII, which comes just before XXV in the ripening series and provides a transitional form to this extreme, all within the essential limits of the type we are considering.

We may sum up the characteristics of the nitrogen effect with apples of classes A and B as being, first, that the respiration lies well above the air-line instead of upon it, and, secondly, that it presents a continuous slow decline. In appearance it may be imagined that the decline, which is faster than that in air, would ultimately bring the nitrogen CO_2 down to equality with the air-line value, such as characterizes the C class, but we have not been able to pursue the problem so far for class A.

The only feature that appears to be common to the two classes is that the first immediate effect of nitrogen is to cause the CO_2 production to rise sharply above the previous level in air. The fact that these two classes should differ in every other particular feature seems to us a striking phenomenon and evidence of a metabolic difference, the general nature of which was only characterized after long analytic study.

On comparing the features of the return to air after nitrogen, it is clear that one essential contrast must exist between the types, arising out of the fact that the A apples have a long way to fall to the air-line while the C apples have been depressed below it, and must rise. Neither of these changes proceeds by what we may call the most direct route. Class C typically shows a sudden transient rise of respiration on return to air, but this subsides again before the final rise to the air-line is achieved. In class A the rapid initial fall is not arrested when the air-line value is first reached, but sweeps below it in a transient dip from which it rises to attain the air-line finally. With apples of classes A and B in the most advanced senescent stage, this typical form undergoes a modification, which suggests an affinity with the type of class C, in that a transient hump of CO_2 production is the initial reaction to air, and this phenomenon is, as it were, superposed on the fall from nitrogen values to the air-line.

3. *The nitrogen effect after gas mixtures other than air*

With all the nitrogen records considered in the previous sections, the apple was respiring in air up to the change into nitrogen. We have now to consider the two cases in which the apple was in an atmosphere either much richer in oxygen ($100\% \ O_2$, apple XXIII) or much poorer in oxygen ($5\% \ O_2$, apple XI). Examination of these records will show what effect this difference of antecedent conditions has upon the respiration in the nitrogen period. Both apples were returned to air at the end of the period in nitrogen. In fig. 2 the records of the nitrogen experience of these two apples have been brought together, alined at the transition to nitrogen.

Apple XI, transition 5 % O_2 *to nitrogen.* The apple was in air up to hour 24, which locates the first stage of its air-line, but was then treated with 5 % O_2. The effects of this gas will be dealt with fully in later papers in this book; it suffices here to observe that the CO_2 production steadily falls in 5 % O_2 till at hour 67 it lies at about 0·7 of the air-line value. At this hour pure nitrogen is given, and we see the respiration rising steadily, cutting the air-line in about 12 hr. and continuing to rise well above the air-line in a way which is characteristic of class A and not

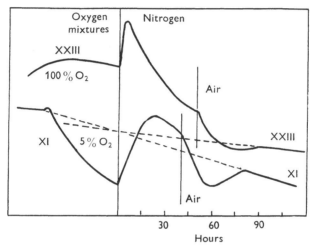

Fig. 2. Two examples of the nitrogen effect after gases other than air, XXIII from 100 % O_2 and XI from 5 % O_2. Both were returned to air after nitrogen. The whole of the CO_2 record observed is given by the heavy continuous line; the broken lines are the air-lines of the two apples.

of class C. The maximum value is not reached till 24 hr., and then there sets in the steady decline of the A type, converging on the air-line. The course of the fall is not well established, as owing to failure of clockwork six 3 hr. readings are merged into one of 18 hr. At hour 110, after 40 hr. in nitrogen, air was given, and the course will be commented on shortly.

Apple XXIII, transition 100 % O_2 *to nitrogen.* This apple was initially in air for 52 hr., so that its air-line is well defined. At this hour pure oxygen was given, and we see the respiration rising steadily to a level far above the air-line and then maintaining a line 1·40 times the value in air.* At hour 124 nitrogen was given, and in spite of the high air value, the respiration rises, at first still higher, but then declines steadily, as is characteristic of an apple of the A or B class. At hour 174, after 50 hr. in nitrogen, it is still markedly above the air-line. At this hour air was given.

* The effect of oxygen upon respiration will be expounded in later papers in this book.

From this survey of the records there seems no doubt that apples XI and XXIII both belong to the type of classes A and B, and neither to class C.

Transition nitrogen to air, records XI and XXIII. With apple XI air was substituted for nitrogen at hour 110, and the form of the record seems to be quite typical of class A. There is an initial big dip below the air-line, and presently the record rises again to join the air-line. Apple XXIII, however, has a different form. When given air at hour 174, the respiration falls smoothly and slowly on to the air-line, showing only a slight dip below it. Our interpretation of this form is that it is to be affiliated to that of XXV as an outcome of the late senescent stage. Apples XXI, XXIII and XXV make a good series of drifting forms in this progress, in which XXIII is clearly intermediate between XXI and XXV, for while the latest case, XXV, rises initially before making for the air-line and XXI dips below it initially, we find XXIII drops moderately on coming into air and reaches the air-line slowly, with only a slight dip in the record.

4. *The nitrogen effect with a mycelium-infected apple*

Apple VIII has a section to itself, because it was the only case in which a patch of mycelium developed on an apple during the course of experimentation. The record of VIII in Appendix I shows that after 60 hr., though still in air, the CO_2 output started to rise at a very rapid rate, quite unlike any other apple of the whole set. The respiration also became very irregular, fluctuating widely from reading to reading, so that the 'contour lines' that contain the whole series have to be widely separated. This state of things could only be attributed to fungal development.

Having carried on the record in air for 160 hr., it was thought worth while to give nitrogen, and see what happened when such active mycelium is present. A sweeping fall of respiration sets in from the high level of 22·8 and continues for 36 hr., finishing with a nearly level series at 9·3. After 12 hr. more of nitrogen, air was given again at hour 210. On this the respiration rose to a somewhat higher level (probably fluctuating), and after 36 hr. fell slowly to the level of 9·3 that it had given in nitrogen. The whole record is so different from others that we can only offer a conjectural interpretation of its features. One thing is clear, namely, that nitrogen cuts out entirely the source of the large irregular output of CO_2 in air. This must turn on the killing of the mycelium by the absence of oxygen, for when air is readmitted there is no recurrence of the high values nor even an upward tendency in the respiration within 80 hr. We can, therefore, use the values after administration of nitrogen, as pure apple values, those before being apple *plus* mycelium. The value 9·3 suggests itself as the air-line value for the later part of the record, as it is the final value in air and also holds for the last 12 hr. in nitrogen. The air-line for the early part of the experiment lies higher and is assumed to begin from an initial value of 12·2, due allowance having been made for the 'initial rise of temperature effect'.

All things considered together, we regard this apple VIII as belonging to the same type as VII (the late ripening group C) and characterized by slowly rising respiration initially, under our standard conditions of experimentation. The air-line is accordingly shown rising slowly from 12·2 to 13·4 at hour 60. It may be that this early rise is contributed to by mycelium development, but we assume it due to the apple metabolism up to hour 60. After that hour the air-line of the apple proper has to get to the lower value of 9·3, adopted for the later part. The falling curve forming the middle of the air-line is purely hypothetical, and its considerable fall implies that a number of apple cells are killed to form the brown patch revealed at the finish, in which the mycelium is at work. Were it assumed that mycelium had been at work unobserved from the beginning, then there need not have occurred any rise at all in apple respiration proper, and the necessary fall to the value of 9·3 in the air-line might be quite inconsiderable.

We return to the phenomena in nitrogen and note that this gas was only continued for 48 hr. A glance at the nitrogen record for VII will show that no depression should be produced in this time and the respiration would be expected to lie on the air-line for hours 20–48 in nitrogen. This is the state of things we have assumed in the air-line here drawn.

Finally, with regard to the nitrogen after-effect when back in air, we note that in neither of the other cases of group C was air readmitted before nitrogen depression (the third phase) had set in. We have, then, a new case here, and we may accept this difference as the explanation of the fact that none of the after-values in air lies below the air-line. The fluctuating form of the after-effect may well be similar to those of VII and XXII, though this is not proved, as owing to the stoppage of the clockwork 21 hr. of readings were merged into one average value, as seen in the record. Anyhow, there is undoubtedly a sharp rise on passing from nitrogen to air, which is a character of group C. Also, it seems clear that after 70 hr. in air, the after-effect of so short an exposure to nitrogen will have completely subsided, so that the value of 9·3 attained may be safely taken to represent the true air-line value.

All the characters of this curious record, VIII, can thus be fairly interpreted in terms of other experiments if it is regarded as being, like VII, a representative of group C.

Summary of Part I

The most striking general outcome of our survey of the ten nitrogen records is that they all belong to one or other of two types, which appear so distinct that there is never any ambiguity about assigning a case to its type. Looking at them from the traditional point of view, which stresses the relation between the magnitudes of CO_2 production in nitrogen and in air, the types seem to have practically nothing in common. It is the purpose of our analytic treatment to determine what is really the common measure of metabolic significance between them.

One most important fact that has been fully established is that long exposures to nitrogen have no permanent disturbing influence upon the metabolism, except in the case of the late class C apple XXIV, for when returned to air the respiration recovers and returns to the same line of starvation drift that it was travelling along before the nitrogen was given. This line is drawn through all the records as the 'respiration air-line', and it clearly expresses some fundamental and dominant ontogenetic drift the march of which is neither accelerated nor retarded to an appreciable extent by exposure to nitrogen, however violent the temporary change of CO_2 production may be. The air-line respiration will provide us with a very stable standard of reference for purposes of analysis.

A second important feature of the nitrogen records is that such striking things happen at the transitions, when the sudden change of gas takes place. Recovery of position on the air-line after nitrogen may take a couple of days and be associated with a temporary phase of very low CO_2 production. Entry into nitrogen, on the contrary, displays only a short transition, and the form is very different with the two types of effects. We hope to elucidate the significance of these transitional forms by the further analysis that we now enter upon.

PART II. ANALYSIS OF THE NITROGEN EFFECTS AS QUANTITATIVE VARIATIONS OF CO_2 PRODUCTION FROM THE AIR STANDARD

Our general outlook upon the apples under investigation has been to regard them as individuals which are undergoing a steady metabolic drift in storage. The broad respiratory expression of this drift is that in its early stages the respiration, as measured at 22° C., may keep level for a considerable time, while at later stages the respiration steadily falls off in a starvation curve.

These phenomena were fully set out in the first paper of this series. The only index of metabolic state and drift that was employed in that analysis was the pitch and the form of the 'air-line' of respiration. Apples of class C, which progressed slowly through the ripening drift, provided examples of the level air-line, while those of classes A and B gave numerous examples of the falling air-line.

The departures of the magnitude of CO_2 production in nitrogen from these air standards might be determined in various ways, and we have to examine all possibilities in a search for fundamental causes. As CO_2 production rates in nitrogen and air tend to be either both at a high level or both at a low level, one obvious inquiry is as to whether there is not some significant *ratio* between the intensities or rates of CO_2 production in the two states. This is the aspect that we may examine first.

1. *A survey of the ratios of CO_2 production in air and nitrogen*

As the air-line of respiration provides us with a stable standard of reference, it is a simple matter to express the CO_2 production in nitrogen or any oxygen mixture as a ratio to the contemporary value on the air-line, which latter gives us the value that would have been produced in air had the apple remained in air all along.

In drawing up our ratios we shall avoid committing ourselves to the use of established terms, like anaerobic and aerobic respiration, and employ the following symbols, which simply refer to actual data of this investigation and the conditions in which they are obtained:

NR (nitrogen respiration) values signify rates of CO_2 production in pure nitrogen.

OR (oxygen respiration) values signify rates of CO_2 production in air or other oxygen mixtures.

ALR (air-line respiration) values are CO_2 values obtained from the air-line drawn through each record.

TR (total respiration) values are CO_2 values without implication as to their metabolic source.

The expression TR is of value at transitions and in cases that will come up for consideration later, where, in very low oxygen supply, we have mixed effects so that TR is useful for the sum of OR + NR. With regard to the relation between OR and ALR, it is clear that when the apple is in air and the respiration is properly adjusted to air, then OR and ALR are identical. The chief use of ALR values is for the periods when the apple is not in air and the air-line is drawn by inter-polation. When an apple is in pure oxygen or $5\% \ O_2$, then the CO_2 production is very different from that in air, and here OR values depart widely from ALR values.

The present section of our analysis is concerned with the variations and drift of the ratio

$$\frac{\text{NR}}{\text{ALR}} = \frac{CO_2 \text{ value (per 300 g.-hr.) in nitrogen, at any point of time}}{CO_2 \text{ value of air-line at the same point of time}}.$$

In table 1 are set out the values of NR, ALR and their ratios which our data provide us with. Nine cases out of the ten described in Part I appear in this table, apple VIII, which was attacked by mycelium, being omitted. Against two it is noted that they came into nitrogen from 100 and $5\% \ O_2$ respectively, while the other seven were in air before nitrogen. The values are set out for points of time, 10 hr. apart, from the beginning of nitrogen on to the end of it, which varies from 40 to 100 hr. The ALR values have been taken from the smooth ALR line, which figures in each record. The courses of observed NR in the original records do not furnish very smooth curves. It has been mentioned already that in nitrogen the actual record for apples of class A fluctuates in a wavy course, and the higher the pitch of the record the more ample are the fluctuations. For analysis by ratios we need a smoother line than some records give directly, and the values of NR set out in the table are smoothed values. There are, however, only two cases where the smoothing we have adopted alters the observed course in a material way.

The cases which have been altered in form are those of XIX and XXIa, the difference between observed and smoothed values being set out in fig. 3. In this figure the observed points are derived from the full records in Appendix I by taking the middle value between the 'contour lines' at the specified hour. The assumption started from in smoothing has been that, as these two apples both belong to class A and are adjacent in date and in condition, they should ideally have a common form of falling NR curve, and that departures from it should be regarded as individual fluctuations. The mean form adopted treats the high values of XIX at hours 20 and 30 as being excessive and the low values at hours 40–60 as defective. With XXIa, on the contrary, hours 40–50 give excessive values. In support of the expectation of similarity is the fact that the air-lines of these two apples have identical forms. Adopting the common form of NR curve drawn in this figure and giving it the appropriate pitch for these two apples, we arrive at the smoothed series of values that appear in table 1.

Table 1. *Table of NR and ALR values and their ratios*

	Hours in nitrogen ...		0	10	20	30	40	50	60	70	80	90	100
C	VII	NR	(20·4)	16·4	13·7	13·7	13·7	13·4	12·2	11·1			
		ALR	13·6	13·7	13·7	13·7	13·7	13·7	13·7	13·7			
		NR/ALR	1·50	1·20	1·00	1·00	1·00	0·98	0·89	0·81			
C	XXII	NR	(21·3)	16·9	15·9	15·8	14·5	12·7					
		ALR	16·1	16·0	15·9	15·8	15·7	15·6					
		NR/ALR	1·32	1·05	1·00	1·00	0·92	0·81					
C	XXIV	NR	(27·6)	23·0	20·8	17·4	14·5						
		ALR	20·8	20·75	20·7	20·6	20·5						
		NR/ALR	1·33	1·11	1·00	0·84	0·70						
B	XXIII (after oxygen)	NR	(24·0)	21·3	19·1	16·9	15·8	15·0					
		ALR	13·2	12·9	12·7	12·5	12·3	12·1					
		NR/ALR	1·82	1·67	1·52	1·36	1·29	1·23					
B	XXV	NR	(16·6)	15·5	14·5	13·4	12·4	11·5					
		ALR	12·5	12·1	11·8	11·5	11·3	11·1					
		NR/ALR	1·33	1·28	1·23	1·16	1·10	1·04					
A	XIX	NR	(22·0)	20·9	20·0	19·2	18·5	17·9	17·4	17·0			
		ALR	14·2	13·9	13·65	13·4	13·2	13·0	12·8	12·7			
		NR/ALR	1·55	1·50	1·47	1·43	1·40	1·38	1·36	1·34			
A	XXIa	NR	(18·85)	17·8	16·9	16·1	15·4	14·9	[14·45]	[14·05]			
		ALR	12·4	12·1	11·8	11·5	11·3	11·2	11·1	11·0			
		NR/ALR	1·52	1·47	1·43	1·40	1·36	1·33	[1·30]	[1·28]			
A	XXIb	NR	(14·65)	13·7	12·9	12·2	11·6	11·1	10·65	10·25	9·9	9·6	9·35
		ALR	9·7	9·5	9·35	9·2	9·05	8·9	8·8	8·7	8·6	8·5	8·4
		NR/ALR	1·51	1·44	1·38	1·32	1·28	1·25	1·21	1·18	1·15	1·13	1·11
A	XI (after 5% O₂)	NR	10·9	13·5	16·0	15·6	14·5						
		ALR	14·9	14·5	14·1	13·7	13·3						
		NR/ALR	0·73	0·90	1·13	1·14	1·09						

The calculated NR/ALR ratios set out in table 1 are all plotted against time in fig. 4. There is a group of four in the middle of the figure which drift downward in the same general way; these are the apples brought into nitrogen from air, while it is seen that XXIII, which comes from 100 % O_2, starts with an exceptionally high ratio for NR/ALR but falls steeply, till it joins the middle group after hour 30. Apple XI, on the other hand, which comes into nitrogen from 5 % O_2, starts abnormally low and proceeds to give rising ratios, till it also joins the middle group about hour 25 and then falls off in a similar way. Our first survey of ratios establishes a new and important fact, which is, that the previous oxygen experience completely alters the NR/ALR ratio for at least 24 hr. after entry into nitrogen, but that later on we get ratios that are about the order that we might expect had the apple come from air into nitrogen.

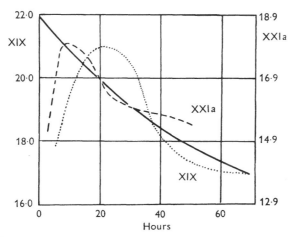

Fig. 3. The smoothed curve of the nitrogen effect adopted for the apples XXIa and XIX, for which the fluctuations observed in nitrogen were very great. The individual records are given as broken lines, the common smooth form of NR as a heavy continuous curve. The pitch of the two records is different, the respective ordinates being given at the two sides.

Coming from an oxygen mixture richer than air we get higher ratios than from air, while lower ratios are obtained when the apple has come from the lower concentration of 5 % O_2. This we register as a significant principle which in itself is in opposition to the idea of the whole situation being dominated by constant ratio relations. Even setting aside these two special cases of XXIII and XI, we find little support for any constancy of ratio. The table shows us that if we had to make a general statement for the behaviour of these apples in nitrogen after air, we could only say that the ratios observed ranged between 1·55 and 1·00.

There is, however, one set of ratios, which we may call the 'initial NR/ALR ratios', which exhibits a certain uniformity. It will be seen in the table that we have given in brackets an initial value for the NR series at the zero hour of entry into nitrogen. This is arrived at by graphic extrapolation of the observed series of NR values set out in the table. When the apple passes into nitrogen from air, the CO_2 production rises considerably and fairly quickly, but it takes some hours to load up the tissues of the apple with the new higher CO_2 content, and it is usually 7–10 hr. before the rising transition brings the CO_2 escape up to the falling line, on which all subsequent NR values lie. The number of points available gives the extrapolation a fairly obvious course, and so we regard these initial NR values as a close measure of the values that would be given if the air supply could be cut off instantaneously and the new higher CO_2 content were to require negligible time for adjustment. As these initial values provide in each case the maximal value and maximal ratio for the function NR, they are obviously of considerable metabolic significance.

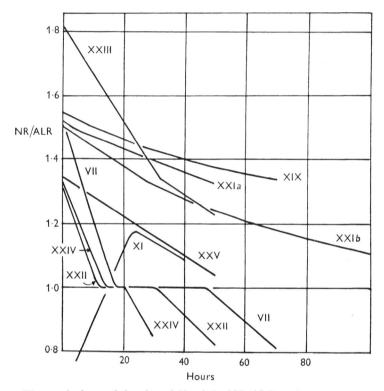

Fig. 4. A chart of the time drift of the NR/ALR ratios set out in table 1.
Continuous curves based on the values calculated for 10 hr. intervals.

Let us then examine the initial NR/ALR ratios class by class. For class A we have cases XIX, XXIa and XXIb, all coming from air to nitrogen, and we find the ratios 1·55, 1·52, 1·51, which from our present point of view we may regard as practically identical. Since we know that XXIb was carried out on the same apple as XXIa, with an interval of 140 hr. in air between, it is striking to find the same initial ratio for both. This demonstrates that the initial ratio is not a relation which shifts with time for a given apple as the metabolic drift progresses. Even though the series of ratios had fallen to 1·33 at the end of XXIa, yet, after recovery of the apple on to its proper air-line, the same initial ratio is given on a second nitrogen treatment; and this, though the absolute pitch of respiration had fallen considerably.

Turning to the apples of class B we have only the case of XXV coming into nitrogen from air. Here the initial ratio is 1·33. In class C we have three cases: the very early case of VII gave 1·50, while the two late cases XXII and XXIV give the ratio value of 1·32 and 1·33. It thus appears as if the initial ratios after air group round two distinct values, which we may round off as the values of 1·50 and 1·33. All the 1·33 relations observed come in apples at the end of the series, such as XXII, XXIV and XXV, while the earlier give 1·50 ratios; but as the classes are not uniformly distributed along the series, it may yet be held that 1·5 characterizes class A, 1·33 class B, while class C shows 1·5 at the beginning and 1·33 values 6 months later.

When we turn from this evidence of uniformity in initial ratios to consider the falling series of ratios for the successive hours, as set out in table 1 and fig. 4, we find no signs of any simple system. The two apples of class B show a more rapid fall of ratios with time than the three of class A, but even within this last group, though the initial ratios are close together at 1·5, the falling series diverge considerably. Nor on this extended time scale is it easy to form any conclusion as to the ultimate value of these ratios had each nitrogen experience been kept on for many days longer.

We may therefore present our data in another graphic way, in which the indefinite extension of the time axis is telescoped. Fig. 5 enables one to inspect the drift of each type of respiration as well as the ratios between them. Here the NR values are plotted against the corresponding ALR values, the two axes NR and ALR having their common origin at zero values at the bottom right-hand corner. Each pair of values from the table gives a point on this co-ordinate system, and the points for each individual are connected up with lines which combine them into a drifting series. The initial point of each apple is distinguished by a circle.

As both forms of respiration are diminishing by starvation, we should expect on any simple scheme that each apple would begin somewhere towards the top left-hand corner of the figure and drift towards the final extinction of respiration represented by the origin at the lower right-hand corner. If NR and ALR should maintain a constant ratio throughout 'the whole progress, then the successive points would, of course, lie on a straight line passing through the origin. A few such 'constant-ratio lines' have been drawn in the figure as construction lines; the constant ratios they stand for are indicated at the left above. There are lines for the following ratios: 1·00, 1·30, 1·34, 1·52 and 1·55. It is at once clear that the co-ordinated values of NR and ALR show no sign of drifting towards zero values in a direct line. They move across the figure to the right, looking at first sight as if they would in time cut the ALR axis high up and provide the strange situation of considerable ALR values in association with zero values of NR.

We may look for a moment at the individual cases in this figure. Cases XXI*a* and XXI*b* are of great interest because they represent two successive exposures to nitrogen with the same apple. Apple XXI*a* has its initial fairly towards the left and lying on the 1·5 ratio track. With successive hours up to the end of 50 hr. the points move diagonally across, almost in a straight course, and reach a value on the 1·34 ratio line. During this period the movement to the right indicates

a drop of 4·0 (18·9 to 14·9) (see the table), while the movement vertically downwards indicates a drop in ALR of 1·2 (12·4 to 11·2). This is less than one-third
of the NR drop. Between $XXIa$ and $XXIb$ there intervenes a period of about
140 hr. in air, but when we start nitrogen again with the initial point of $XXIb$, we
find that drop of ALR in this air period has been 1·5 (11·2 less 9·7), while the shift
to the right, i.e. along the NR axis, has only been 0·2 (14·9 less 14·7). This position
allows us to formulate the statement that, when actually in nitrogen, both NR
and ALR values fall off (the former about three times as fast), but when in air,
though ALR continues to fall off, yet NR hardly falls off at all.

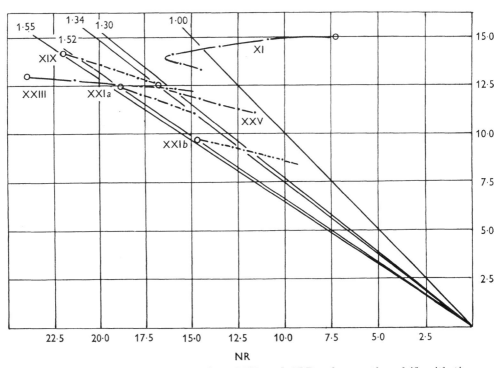

Fig. 5. A chart showing the co-ordination of NR and ALR values as they drift with time.
The axes are values of NR and ALR with a common origin at zero. Five construction lines
for certain constant ratios for NR/ALR are drawn, and the ratio value for each line is inserted
at the top left-hand corner. The co-ordinated values of NR and ALR for all the class A
and B apples are inserted in the chart by a series of dots for each 10 hr. interval, as given
in table 1. The points of each individual apple are linked together by a heavy broken line,
the initial value being distinguished by a circle.

It is therefore made obvious that analysis by series of ratios between NR and
the ALR values is not going to clear up the true situation. The one ratio that
appears to have a real significance is the 1·5 ratio, which reappears at the initial
of $XXIb$, still the same as the ratio at the initial of $XXIa$. Though both respiration
values of $XXIb$ are lower absolutely than at the beginning of $XXIa$, yet the ratio
of them is maintained. This suggests a simple form of experimental relation,
namely, that if one continued to alternate nitrogen and air during the starvation
progress towards zero at the origin, every one of the initial points on passing into

nitrogen would lie on the 1·5 ratio line, which makes directly towards the origin. This we may hold fast to, as being a significant fundamental relation.

With regard to the drift of NR and ALR values during the nitrogen experiences of XXI*b*, we see a repetition of the rapid drift to the right, so that after 50 hr. NR has dropped 3·6 (14·7 less 11·1) and ALR 0·8 (9·7 less 8·9), the former 4·5 fold the latter, so that the line of drift is more horizontal than in XXI*a*. During the further 45 hr. in nitrogen the values show signs of working downwards. Possibly this downward drift would become more marked had nitrogen been continued longer, so that presently the value would have got on to the 1·0 ratio construction line. This is merely a formal suggestion at present; we see that the points on the track of XXI*b*, representing 10 hr. periods, are already getting very close together, forecasting perhaps another 100 hr. in nitrogen to attain the 1·0 ratio construction line.

Passing now to the other apple of class A, namely, XIX, which was earlier in chronology than XXI*a*, and has therefore higher ALR values, we find that it has also higher NR values and that its initial point has again the 1·5 ratio relation, so that by this and by its drift of values while in nitrogen it takes its place on the chart quite as if it were an earlier nitrogen experience of apple XXI.

When we consider the only other case of transference direct from air to nitrogen, that of the class B apple XXV, we note that its initial ALR has about the same value as XXI*a*, but its NR value is less, so that the initial point lies on another construction line, that of the 1·33 ratio. Kept in nitrogen it shows exactly the same type of behaviour as the class A apples, drifting rapidly to the right, NR having dropped in 50 hr. 5·1 (16·6 less 11·5), while ALR has dropped 1·4 (12·5 less 11·1), nearly a fourfold drop in NR. This drift brings it by this short time nearly on to the 1·0 ratio line and it is still drifting rapidly. So that we conclude that though a succession of initials in nitrogen would, for this B apple, fall always on the 1·33 construction line, yet long-continuous nitrogen would probably drift the values across the 1·0 ratio line sooner than in class A.

The other class B apple, XXIII, shows the effect of exposure to high oxygen concentration before nitrogen in that the nitrogen initial has a much higher NR value than any other apple, though the ALR value is much lower than XIX and little higher than XXV, the ratio here being as high as 1·82. It recovers from this temporary effect after 25 hr., and its three later values are so placed that they mingle with the points of XIX and XXV.

Apple XI shows the opposite displacement from XXIII. Its initial ALR value is the highest of the lot, being an earlier class A apple, but as it has been in 5 % O_2 before nitrogen, its early NR values are very low and rising rapidly while ALR falls. These are special transitional values, and the first three points might well have been omitted from this graph. Soon after the fourth point, this transition is over and its maximum NR value of 16·4 is reached. Now at last the record begins to behave like the other cases and shows values drifting across to the right and a little downward. The track is such that if nitrogen had been continued longer the values might soon have drifted on to, and across, the 1·0 ratio construction line. This apple is the only early class A apple investigated in nitrogen, and its case is difficult because we have no example of this particular metabolic state which had been brought into nitrogen directly from air. We return to this case in § 3 of Paper III after further analysis has been carried out.

None of the class C values has been entered in this figure because it is clear that, since ALR values remain practically constant throughout their records, the series of NR points will be merely a procession across the graph from left to right on one horizontal line, and only the distances between the points, indicating periods of 10 hr., will have any significance. Until we come to investigate apples so early in storage that their ALR line is definitely rising, class C furnishes the extreme type, giving points on a horizontal line, whereas all the A and B apples lie on a line falling somewhat away to the right. The slopes of these A and B lines are nearly all about parallel, moving 3–4 units to the right for a drop of 1 unit vertically, indicating that NR is falling off three or four times as quickly as is ALR.

The absence of any tendency for the series of experimental points in this figure to move towards the common origin during the fairly long periods that the apples were under examination suggests some essential divergence of the course of NR from that of ALR. The drift apart of the two values is of the type that would be expected if NR were drifting towards zero at a relatively rapid rate, while ALR drifted in that direction at a very much slower rate. The striking thing is that recovery of the ALR position is complete on going back to air, so this tendency does not persist in air after nitrogen.

If an apple, after being kept in nitrogen for 100 hr., could not be brought back to its ALR values by subsequent air, but was found to be permanently depressed, then it would be assumed that some toxic effect had inactivated the machinery, but of this there is no sign, and we regard the depression as metabolic. The only possible type of toxic effect that would fit the facts would be an auto-depression of NR by some product, such as alcohol or acetaldehyde, that accumulated in the tissues, combined with the absence of depression of the ALR values subsequently. Instead of complete absence of depression of ALR subsequent to NR, it would be possible to hold that the toxic substance was rapidly oxidized to an innocuous one in the early hours of the return of the oxygen supply.

In this connexion we have calculated the expected alcohol content of the apple tissues at the end of several of the nitrogen experiences, on the basis that 1 molecule of alcohol is produced in association with each molecule of CO_2. This works out at 1·045 mg. alcohol added to the tissues for each 1·0 mg. CO_2 escaping during the progress of NR. In the 70 hr. of XIX, 439 mg. CO_2 escape, and therefore 457 mg. alcohol are accumulated, giving a concentration in the tissues of 0·45 %. In the 100 hr. of XXI*b*, the totals per 100 g. of apple are respectively 369 mg. CO_2 and 384 mg. alcohol, giving a concentration of 0·38 % alcohol. Direct experiments have yet to be tried upon the effects of small additions of alcohol. No allowance for any such depressant effect has been made in the present survey.

By our inspection of NR/ALR ratios we have been led to attribute a certain amount of individuality to NR behaviour by which it departs from ALR behaviour. We have also established the definiteness of initial NR/ALR ratios after air; and, further, that quite different ratios are obtained when the apple has been brought from higher or lower concentrations of oxygen. We may now take up another line of analysis.

2. *Analysis of the relation of ALR and NR drift by difference of values*

The uniformity underlying the relation of a pair of falling curves, if not to be found in their ratios, might possibly reside in their arithmetical differences, and when we turn to investigate this aspect of the pairs of ALR–NR curves, we come across a significant relation. The only region where we have several suitable cases available for comparison is the group of three records XIX, XXIa and XXIb, all of which are class A and closely adjacent in date and condition.

In all these cases NR is greater than ALR, and the differences, NR less ALR, as derived from the values set out in table 1, p. 42, are brought together in the upper part of table 2. In the column for zero hour, where the ratio NR/ALR is, for class A, about 1·5, we find a decreasing set of differences, in the serial order of the cases, each difference being, of course, half ALR. Similarly, the differences decrease in the column for each other hour. Inspection of the whole set brings out that all the series of differences run a parallel course. The series of XXIa is always 1·3 to 1·2 below that for XIX, and the series for XXIb is always 2·8 to 2·75 below XIX. The sets of values are characterized then by constant differences.* The record for XXIa actually stopped at hour 50, and the two end-values in brackets are additions calculated on the basis of constant differences, so that we may obtain three sets of 70 hr. duration for further examination.

Table 2. *Table of differences NR less ALR*

(For values of NR and ALR, see table 1, p. 42)

Hours in nitrogen	0	10	20	30	40	50	60	70
Class A, XIX	7·8	7·0	6·35	5·8	5·3	4·9	4·6	4·3
XXIa	6·5	5·7	5·1	4·6	4·1	3·7	[3·35]	[3·05]
XXIb	5·0	4·2	3·55	3·0	2·55	2·2	1·85	1·55
Class B, XXIII	10·8	8·4	6·4	4·4	3·5	2·9		
XXV	4·1	3·4	2·7	1·9	1·1	0·4		
Class C, VII	6·8	2·7	0·0	0·0	0·0	− 0·3	− 1·5	− 2·6
XXII	5·2	0·9	0·0	0·0	− 1·2	− 2·9		
XXIV	6·8	2·25	0·1	− 3·2	− 6·0			
Class A, XI	− 4·0	− 1·0	+ 1·9	1·9	1·2			

In fig. 6 we present all the NR curves in terms of their excess values over their companion ALR curves. Those of class A, now under discussion, are there grouped at the bottom. It is clear that the more starved the apple becomes, giving lower

* It may not escape the reader's notice that contribution has already been made to the parallelism of XIX and XXIa by the procedure adopted for smoothing the NR curves for these two cases carried out in fig. 3. Had not the ALR values also run a parallel course for the two apples, then the parallelism of the differences of the pair would not have resulted, so that the two results do give more than a single measure of support to the independent case of XXIb.

ALR values, the smaller will be the initial excess of NR if the NR/ALR ratio remains constant. Since the NR curves are all parallel, the smaller also will be the excess of NR at each successive hour. To the three upper observed curves of this group we have added an imaginary lower curve also running a parallel course. This gives the curve of NR that we should expect on these principles when ALR falls to 6·9, and therefore the excess NR over ALR becomes initially half this value, 3·45. This curve has been selected as one that would just cut across the ALR curve at hour 70. After that hour the NR values will be below the ALR.

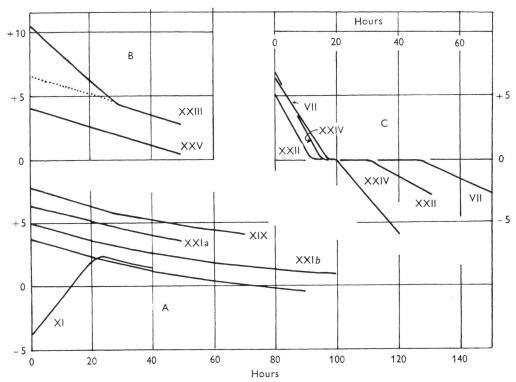

Fig. 6. A chart of the drift of differences, NR less ALR, with time in nitrogen. The three classes of apples A, B and C are given in separate groups. In class A, beside the four numbered curves which run a parallel course, an additional curve has been constructed on the same basis, but of lower initial. This unnumbered curve cuts the air-line at hour 70. In class B a dotted addition has been made to the observed curve of XXIII which came to nitrogen from pure oxygen. This constitutes a suggestion of the form that might be expected had XXIII been brought into nitrogen from air instead of oxygen. The group of class C curves is described on p. 53.

How much further all these NR curves would drop after hour 70 cannot yet be safely predicted. The curve for XXI*b* was followed on to 100 hr. and continued to fall and approach the ALR value. Graphed on this scale it looks as if it might flatten out and never reach ALR, but plotting on a more condensed scale suggests arrival there in 170–200 hr.

In order to present the relation that we have just made out for this group of curves in a more realistic setting, we have drawn up fig. 7, in which the three cases examined are set out on one time axis. As XXI*a* and XXI*b* were carried out on the same apple with a known time in air between them, these two cases can be accurately spaced on the time axis. The case of XIX preceded XXI, but as it was a separate individual, the two time scales cannot be combined into one accurately. We have therefore located XIX in front of XXI*a* to such an extent as will allow its air-line to run on into the line of XXI*a* in a natural falling curve, just as if all three cases had been carried out on one apple. The general disposition of the parts of this figure can be checked on the detailed records in Appendix I.

The heavy line right through the figure gives the absolute ALR values and the three falling NR curves are also marked in heavy lines. As construction lines we have connected up the three initial NR values by what we may call an 'initial nitrogen line'. This line could have been drawn by eye well enough, but the values were calculated from the known ALR values in the three cases by using the factors established for the three actual initial NR values; these factors being, as table 1 shows, 1·55 ALR for XIX, 1·52 ALR for XXI*a* and 1·51 ALR for XXI*b*. We have added further a line connecting up the NR values observed at hour 30 for the three cases and another line for the values at hour 70. We see that the '30 hr. nitrogen line' and the '70 hr. nitrogen line' are practically parallel to the 'initial nitrogen line'.

The relation of this system of nitrogen lines to the ALR line now becomes clear. Since the initial nitrogen line has an approximately constant ratio (1·5) to the ALR line, then the other NR lines parallel to it cannot have a constant ratio to this same line. They must tend to cut the ALR line in serial order at points of time which are earlier, the lower the absolute value of ALR.

The lowest NR line we have drawn is the '70 hr. line'; the 100 hr. line would be lower still, but only XXI*b* provides us with a point for this line. Uncertainty as to the form of the continuation of the NR lines after hour 100 prevents us from completing the schema, but we have suggested that these lines continue as falling straight lines at the slope that hours 90–100 indicate.

As an inset at the bottom of the figure we have added an imaginary case constructed on the principles set out above. Here ALR is supposed to have fallen as low as 5·0, and the initial NR will therefore be 5·0 × 1·5. The initial NR line is drawn at this ratio to ALR and the NR sequence is drawn falling away by the same series of values as in the earlier cases. As a result it is clear that in this extreme case the NR line cuts across ALR as early as hour 42, and thenceforward diverges from it below.

By this survey of NR and ALR relations we have established a new and significant principle, which is that the grade of starvation is the thing that determines whether the course of NR, when the initial hours are passed, lies well above the air respiration, or close to it, or mostly below the air-line. All these states might possibly be produced in one individual apple, allowing, of course, for the fundamental relation in class A apples, that the initial NR is 1·5 times the contemporary ALR value. This point of view stresses still further the part played by carbohydrate metabolism in determining relations which might *a priori* have been attributed to oxidation processes.

Before going further into this aspect of respiration we have yet to inspect our other NR curves and see what indications of regularity based on values of NR less ALR they may present. There await us the two cases of class B, XXIII and XXV, the three cases of class C, VII, XXII and XXIV, and the single early case of class A represented by XI. In table 2 the values are set out and they are presented graphically in fig. 6, groups B and C.

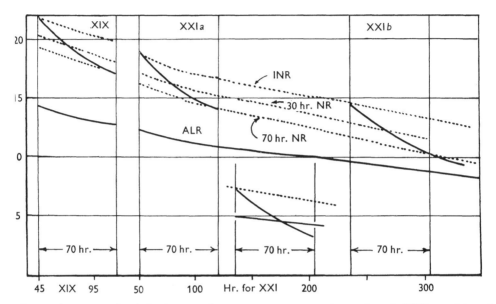

Fig. 7. A construction bringing together the two nitrogen effects of apple XXI*a* and *b* and that of XIX for comparison of the relation of drift of NR to drift ALR. The two parts of XXI are correctly spaced on the actual time axis of the record, which is given below, but the case of XIX is arbitrarily placed on this time axis, so that its ALR line runs on smoothly into that of XXI. The heavy continuous curves above represent the course of downward drift of NR in the three cases between vertical lines marking off the first 70 hr. in nitrogen. With XXI*b* the NR curve was observed longer, on to 96 hr., but with XXI*a* the actual observed period was only 50 hr., the values for hours 60 and 70 being calculated as shown in table 2.

Three construction dotted lines are added to the nitrogen part of the record. These join up for the three cases (1) the initial NR values at zero hour, (2) the values at 30 hr. in nitrogen, and (3) the values at 70 hr. in nitrogen. These three nitrogen lines are practically parallel. It is brought out in this way that the lower the absolute pitch of NR initially the sooner will the continuation of the NR line make contact with the ALR line.

At the bottom of the figure in the middle an imaginary case has been drawn up, where ALR is only 5·0 when nitrogen is given. On adding to this an NR curve, constructed on the principles set out in the text, we find that NR cuts ALR as early as hour 40.

We may look first at the two cases of class B. Apple XXV brought from air to nitrogen distinguished itself from class A by showing a steep and rectilinear fall of NR/ALR ratios. The values of excess NR over ALR also fall off steeply in a fairly straight line, so that in 50 hr., when nitrogen was stopped, the difference had nearly become zero. The initial excess of NR is 4·1, giving an NR/ALR ratio of 1·32 instead of the 1·5 ratio of class A. There are thus three minor points distinguishing XXV from the group XIX–XXI*b*, namely, the steeper fall, the linear course, the early prospective crossing of the ALR line, to add to the different initial ratio.

Grouped with XXV in fig. 6 there is also the curve of values of differences for XXIII, another apple of class B. As this apple was in pure oxygen before nitrogen, its OR value was much above the ALR, and the effect of this previous history was to give a very high initial NR with an ALR ratio of 1·82. The excess over ALR is initially 10·8, and the values fall off very steeply in a straight line till hour 30, after which the fall continues much less steeply on to hour 50, when the nitrogen was ended. In this last part the line of XXIII runs about parallel to that of XXV; but in the first 30 hr. the high values are due to the previous exposure to oxygen. We can imagine that if XXIII had come into nitrogen from air, the series of NR − ALR differences would have run parallel to that of XXV all along, and a dotted line has been drawn in the figure for hours 0–30 on that assumption. This indicates for this hypothetical treatment of XXIII an initial NR − ALR difference of 6·6. The recorded ALR value in table 2 is 13·2, so there results on these lines an NR/ALR ratio of 1·50 for this apple going into nitrogen from air. For the companion apple, XXV, of this class the ratio was shown to be 1·33, so it is left uncertain as to whether this reconstruction is valid as a support of parallelism of difference values. We have no experimental test case to guide us.

Turning to apple XI of class A we have there the opposite distortion from XXIII in that this apple came into nitrogen from 5 % O_2, which had lowered its respiration, and the NR values for some hours are below the ALR line. They rise fairly fast and are well in excess of it by hour 20. After a maximum at hour 25 the values fall off till hour 40, when, too soon for the full information which we now desire, nitrogen was changed for air. The values in table 2 are added to fig. 6, group A, showing that after hour 25 we get the falling curve characteristic of class A and also lying parallel with the rest of the class. If this parallelism between hours 25 and 40 is carried back to zero hour, we get the line, figured, but not numbered, which we might expect to be the course followed had apple XI been in air before its nitrogen experience. The initial value of NR excess over ALR indicated by this graphic construction is about 3·7, while the ALR value has been taken as 14·9. As in the case of XXIII, this graphic method does not give a concordant ratio, indicating NR/ALR = 18·6/14·9 = 1·25.

In no other case have we met a ratio that was not close to either 1·50 or 1·33. We must leave this problem open, after recalling that XI is the only early apple examined in nitrogen for classes A and B, and that special changes of organization-resistance are still proceeding in this early stage, on the lines set out fully in the first paper.

Apples for class C. Finally, we have to survey the special class, C, consisting of apples which show striking characteristics associated with their extremely slow rate of ontogenetic development. The values for the excess NR over ALR are set out in fig. 6, group C. In appearance, this set is widely different from the other two sets, as each case is built up of three jointed linear courses, but nevertheless these lines run parallel for the three different apples. The initial excess is high, but then the ALR values are high in these cases. Inspection of the values in the tables will show that there are here two cases of initial ratio $= 1\cdot33$ and one of $1\cdot55$, while in regard to the $NR - ALR$ differences there are two cases of $6\cdot8$ and one of $5\cdot2$. In spite of this rather mixed state it is clear that the lines are generally parallel in the figure. To this falling phase succeeds the level phase, where $NR/ALR = 1\cdot0$, and here parallelism becomes identity of course. In the third phase, VII and XXII show also parallel movement, while XXIV drops somewhat more steeply. In this last apple there is probably a toxic factor, for the apple succumbed completely with continuation of the nitrogen.

Summary of Part II

The main conclusion that we draw from this survey of NR and ALR drifts is that the metabolic factors that determine the progress with time are different for the two cases. There is a close correlation of the initial magnitude of NR with ALR, but after that the NR values follow a course of their own. For full information about this course we should have to study prolonged starvation drifts in nitrogen in the way we have studied starvation in air. We do not yet know how far this study could be carried with class A and B apples without the interference of toxic effects. One thing is clear, namely, that the grade of starvation in air determines whether the NR values lie above the ALR for hundreds of hours or only for a few tens of hours.

All these considerations lead us to take up the definite position that the difference between the air condition and the nitrogen condition is something other than the difference between CO_2 produced by oxidation of sugar and CO_2 produced by splitting of sugar. The simple ratio relations which might govern a situation determined wholly by oxidation are not found to be applicable. We conclude that a consideration of the carbohydrate metabolism which is antecedent to respiration and supplies the substrate for this catabolic complex must be brought into our field of inquiry.

This extension of the analysis is presented in Paper III.

III

FORMULATION OF A CATALYTIC SYSTEM FOR THE
RESPIRATION OF APPLES AND ITS RELATION TO OXYGEN*

CONTENTS

This paper is an extension of the analytic study started in Paper II, which dealt with the respiration of apples in nitrogen. In Part I of that paper the phenomena of CO_2 production in nitrogen were described empirically, as they presented themselves in our records. Part II was devoted to examination of numerical relations, such as ratios and differences, that could be established between CO_2 production in air and CO_2 production in nitrogen. This present paper is, in effect, Part III of the analysis, but it starts out on new lines, though use is made of various significant ratios established in Part II.

We here attempt a more realistic analysis of the phenomena than was possible when attention was concentrated merely upon CO_2 production. This advance in realism is based upon bringing into our survey the whole drift of the metabolites involved in respiration and picturing this drift as a system of catalysed reactions.

1. THE CATALYTIC SCHEMA PROPOSED FOR THE
RESPIRATION OF APPLES

This system takes the form of a chain of reactions, so that the products formed by one link-reaction become the reactants of the next link. At its free end the chain of reactions is branched and we find alternative fates for reactants, controlled by the oxygen supply.

We have simplified the system as much as possible, but must take account of at least half a dozen catalysed reactions. These will be formulated only for broad schematic treatment, as our correlation of them is merely preliminary to fuller investigation. It will save premature commitments to specific molecular reactants,

* This paper was published in 1928 as the third paper in the series 'Analytic Studies in Plant Respiration'. The author was F. F. Blackman, the reference being *Proc. Roy. Soc.* B, **103**, 491, 1928.

if we represent the substances and stages involved by a formal sequence of letters. Our suggested schema is set out below.

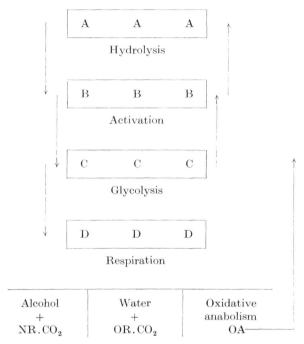

A	A	A

Hydrolysis

B	B	B

Activation

C	C	C

Glycolysis

D	D	D

Respiration

Alcohol + $NR.CO_2$	Water + $OR.CO_2$	Oxidative anabolism OA

Schema of reactions comprising the respiration sequence.

The schema starts with a first group of reactants, entitled A, which includes all the substances in the apple which may function as reserves of carbohydrate and give rise by hydrolysis to free normal hexoses. These normal hexoses constitute our group B. We have assumed in our scheme that these normal hexoses are not respired directly, but that a further carbohydrate stage intervenes, which we may call activation, leading to the formation of hexoses of the group of heterohexoses with the less stable type of internal ring structure. This group, C, is regarded as the direct substrate of the next reaction entitled glycolysis.

Here we come to the well-known activity of the zymase complex leading typically to the formation of alcohol and CO_2. It is known that in the absence of oxygen, apples do actually produce these products abundantly. The study of this process in yeast shows that it is an elaborate complex with many intermediate products with two or three carbon atoms, of the type of methylglyoxal, lactic acid, pyruvic acid and acetaldehyde. For the purpose of our simplification all these intermediate products of glycolysis are grouped together as D, though ultimately it may be possible to differentiate them as D_1, D_2, D_3, etc. Glycolysis here signifies the stage of conversion of reactant C to product D. The products D are assumed to be the reactants for the last stage and so have alternative fates bound up with the presence or absence of oxygen. This last stage we may speak of as respiration in a narrow special sense. In nitrogen, group D proceeds quantitatively to the two final products CO_2 + alcohol in the usual ratio, and these escape from the system as waste end-products. The CO_2 diffuses out and its production rate can be measured; it will be held that for one atom of carbon thus detected there are two atoms of carbon excreted into the tissues as alcohol.

We propose, in this quantitative consideration of the workings of the system, to adopt the carbon atom as unit and so evade specific molecules. When previously we considered respiration of apples in nitrogen merely as a phenomenon of CO_2 production, we termed this production 'nitrogen respiration' 'NR', and we shall maintain this nomenclature, noting that when, for instance, it is stated NR = 1·5, the constant implication will be that this is a measure only of that part of the carbon loss appearing as CO_2. Taking into account the simultaneous production of twice as many carbon atoms in the form of alcohol it is clear that glycolysis = 4·5 is the exact equivalent of NR = 1·5.

In air and other concentrations of oxygen the respiration system behaves otherwise, and one of the main objects of our analysis on the present lines is to attempt to suggest a more precise formulation of the effect of oxygen than has hitherto been put forward. The problems involved will be developed more fully later, after this preliminary sketch of the system is concluded.

It is known that oxygen has no direct effect on zymase activity in yeast, and we shall assume that in apples glycolysis is equally effective in any gas mixture in converting C to products of the group D. In air, however, the only detectable final products escaping from the system are $CO_2 + H_2O$. Now in apples we find that the CO_2 production by this process, which we label as OR (oxygen respiration) is less than the CO_2 production by NR. Clearly then the total carbon loss is three or more times as large in nitrogen as that in air. What then happens to this deficit of carbon in air? No final carbon derivative of D accumulates in the tissues during OR, so the logical conclusion seems to be that in air part of the group D is somehow worked back into the system continuously by oxygen. There is therefore a call for a third reactive mechanism, dealing with D, which we shall speak of as oxidative anabolism, OA. It is round this conception that problems are densest.

This anabolic building back is specific to the presence of oxygen, but of course short-range up-grade reactions occur in those link reactions which are held to be directly reversible. Substances of group B can pass to A by condensation, and C can pass to B by reversion. In contrast with these early stages we shall assume that C → D is not reversible.

Other reactions than those set out are possible, such as the direct origin of C from A, since many reserve carbohydrates contain heterohexoses in their structure. In yeast it seems possible that alcohol may undergo oxidative anabolism to carbohydrate, but the evidence available for apples indicates that alcohol once formed in the tissues remains unalterable.

2. A GENERAL SURVEY OF THE WORKING OF THE PROPOSED RESPIRATION SCHEMA

This system of linked reactions brings together into one sequence processes generally regarded as belonging to separate chapters of physiology, and combines matters of carbohydrate equilibrium in tissues, stages A–B–C, with processes more specifically connected with respiration D–NR–OR. The whole system therefore presents a number of rather complex relationships, and it may be well to give a general survey of its supposed workings before trying to relate it quantitatively to the actual data of our respiration records. This latter aspect will be the subject of the following sections.

(a) The general drift of respiratory activity in starvation

Taken as a whole the schema is intended to represent the fundamental continuous catabolic drift of carbon compounds from the highest molecular structure and energy content towards lower states of these attributes, beginning with production from reserves at the top and ending, at the bottom, with excretion of respiratory waste products.

All along the chain the drift is determined by balance of production and consumption. The controlling factors are the activity of the catalytic components of the system and the amounts of the substrates available. When such a system is studied in a living tissue isolated from the plant and no longer able to increase the amount of A, then we are in the presence of what we may call, broadly, starvation phenomena, but the course of these, when followed in time by measurement of the respiratory products, may appear very different in different tissues. One fundamental variable is the amount of A in relation to the activity of conversion of A to B, and another the *absolute* activity of the final respiratory catalytic system. The apple is distinguished by having a very inactive respiratory system in relation to the amount of A and B, so that the isolated tissue may maintain the system in normal working for many months, or even a year, on its original stock of A.

When the apple is freshly gathered in autumn its respiration may be low and then proceed to rise, in spite of isolation, a phenomenon which we have discussed in the first paper of this series, and attributed to a 'decrease of organization resistance' which is effectively an increase of 'hydrolysis facility' leading to more rapid production of C from A and B. This brings about a rise of respiration and greater excretion of final products out of the system. In the later stages of isolation in storage, to which belong the apples now to be considered, this rise of respiration is over and the whole system exhibits steadily declining magnitudes of catabolic drift and decreasing respiration. This decline is, at 22° C., at first relatively rapid, but it gradually slackens off, appearing to reach in time a slow rectilinear fall. Of the form of such a decline, the air-line of apple XXI*a, b* is taken as a typical example. The average intensity of respiration in this starving case is about 10 mg. CO_2 per 100 g. apple per 3 hr. Calculating this as hexose oxidized per day we find that the loss is only 0·057 g. Assuming that the apples still contain 10 g. hexose per 100 g. tissue, the daily loss is no more than 0·5 % of the stock of B. In addition, there is the production of B from A to compensate this consumption, so that the drift of starvation grade from day to day is very slight and we almost approach a state of dynamic equilibrium. Nevertheless, we must recognize that the system is always progressively starving, and the excretion of carbon at the distal end always somewhat greater than the initial production value, which lags behind it. The steady states of this nature, in which the rate of respiration is determined by production at the top, we propose to distinguish as 'adjusted states' of the system.

Since we have already established that the loss of carbon in nitrogen may start as high as 4·5 times the loss of carbon in air, we see that on change from one gas to the other we do not merely alter the nature of the biochemical mechanism but also we suddenly change the carbon drift quantitatively and alter what we may call the starvation grade. We want, therefore, to find out how the chain and its separate links react to this sudden alteration of consumption at the free end. Realization of this quantitative difference between NR and OR has helped to drive our thoughts back to the production of carbohydrate substrate for respiration, and led to our linking up carbohydrate metabolism and respiration into one production-consumption system.

The sudden change of carbon loss produced by alteration of oxygen supply we view as a sudden disturbance of the carbon traffic system, displacing it from the previous adjusted state. If the new set of conditions is maintained long enough we may expect the system to settle down to a new adjusted state. The relations of the various adjusted states in different conditions of oxygen supply should illuminate the control mechanism of the transport system, as should also the character of the transitional unadjusted states through which the system passes in moving from one adjusted state to another. All these changes are slow in apples because the activity of the catalyst systems is, absolutely, slight. The result of this is that long tedious experimentation is involved in ascertaining the relation of final adjusted states, but there is the compensating advantage that the transitions are of a comfortable slowness so that they can be followed in considerable detail. Tissues provided with a more active catalytic system might pass from one adjusted state to another too quickly for experimental determination of the course of the transition. We shall learn much from the forms of transitions.

(b) The reaction of the respiratory system to alterations of oxygen supply

Having sketched the carbohydrate starvation aspect of the system we may proceed to inquire into the more strictly respiratory processes and the influence of oxygen. It will be proved in the next paper that the intake of O_2 and the output of CO_2 are increased by rising external oxygen concentration. Details are reserved until that exposition, but we need here to realize that if the respiration in air is taken as unity, then that process steadily accelerates with rising oxygen up to a value of 1·4 in pure oxygen and steadily falls off with decrease of oxygen till it reaches a value of 0·7 or less, in 5 % O_2. As we have found in our present study of nitrogen that CO_2 in nitrogen is greater than CO_2 in air, we perceive that there must be a minimum of CO_2 production, though not of oxygen intake, in some one low concentration of oxygen. In apples this occurs in the region of 5% O_2.

In face of these facts, our first tendency was to regard the increase of respiration with increased supply of oxygen as due to increase of the respiratory oxidation by the greater concentration of one of the substrates of respiration, namely, oxygen. More analytical consideration of the data available tends to lead us away from regarding this as the fundamental interpretation. Our present object is, then, to inquire exactly what changes oxygen does produce in the system.

For analytic treatment of the system it seems to us that the long chain of reactions must be divided into three separate short chains, though each will be linked to the next by the fact that its own end-product is the primary reactant of the next stage.

The first stage is the carbohydrate one, A–B–C, concerned in the production of C. The second is the glycolytic system concerned in the production from C of the intermediate products D, and the third is the oxidative system concerned in production of final products OR and OA from D. Unlike the others this last stage can be thrown entirely out of action by nitrogen so that zero activity of it with substitution of NR comes into the possibilities. This being so, we shall provide ourselves with a more general formulation by defining our three stages as (1) glycolysis, (2) stages antecedent to glycolysis, and (3) stages subsequent to glycolysis. This analysis seems to us profitable, because our data lead us to the conclusion that the relation to oxygen is different and characteristic for each of the three stages.

The relation of glycolysis to oxygen. It is well known that oxygen has no direct effect upon the zymatic glycolysis of sugar in yeast, so we should approach this matter with the expectation that the same would be true of glycolysis in the apple. By glycolysis, we understand here the conversion of the substrate C to the products D so that the term has only the narrowest denotation. The first problem is the experimental determination of the magnitude of glycolysis in the different circumstances. In nitrogen the apple produces alcohol as well as CO_2, and if we assume that the proportions of these are the usual ones then it is clear that the measure of glycolysis in nitrogen is, in terms of carbon units, threefold the observed CO_2 in nitrogen. So we can always write for nitrogen, $Gl = 3NR$. For estimation of glycolysis in air or oxygen mixtures we have no such direct measure. As OR in air is found to be much less than NR we might draw the superficial conclusion that oxygen had reduced glycolysis considerably.

On the Pfefferian view of aerobic respiration it would be assumed that in air the whole of the carbon of glycolysis appears as CO_2, while in nitrogen only one-third of the carbon is constituted by CO_2; thus the conclusion that oxygen reduced glycolysis would be inevitable. However, we have satisfied ourselves that oxygen does not have this effect, and we consider that we can employ an experimental procedure which will enable us to arrive at the actual magnitude of glycolysis that is being carried on by an apple in the adjusted state in any oxygen concentration.

The procedure is this: the oxygen current is suddenly replaced by a current of nitrogen and the CO_2 output measured hour after hour. It is found that the CO_2

rapidly rises to a high level and then steadily declines for a long time. This declining series gives a point to point measure of NR and of glycolysis. If now the series of values be extrapolated backwards to the zero hour of entrance into nitrogen, we get a measure of what the glycolysis would have been at zero hour could the transition have been made instantaneously, before starvation had progressed further. This initial value is the value of glycolysis that was going on actually in the adjusted oxygen state previously. We get full confidence in this evaluation of glycolysis when we find that for a given physiological state this initial NR value has, every time, the same ratio to the last OR value at the moment of change into nitrogen. Subsequent behaviour at the transition, on bringing the apple back to air from nitrogen, confirms our estimation of this magnitude of glycolysis in a way that will be brought out clearly in a later section on transitional relations (p. 65). Each 'initial value' of NR in nitrogen that has been under discussion in Part II of the previous paper supplies us then with a measure of the glycolysis in air or the other oxygen mixture that preceded the nitrogen. We thus have at our disposal in that paper, material for evaluations of glycolysis in oxygen mixtures in quite a number of the records, based upon the careful examination of their transitions into and out of nitrogen. These glycolysis records will be set out graphically later in this paper.

This we consider an important advance, as in measuring glycolysis we are measuring the common antecedent stage to both types of respiration and so get the common measure of both NR and OR, or of a mixture of OR + NR. These evaluations of glycolysis establish that glycolysis is really maintained at a higher rate in air than it can be maintained in nitrogen; and further, that glycolysis is 1·4 times as high in pure oxygen as in air and is depressed to 0·7 of the air value in 5 % O_2. The facts show that glycolysis moves up and down with oxygen concentration just as OR does. The magnitude of glycolysis relatively to OR is high, so that for some classes of apples Gl = 4·5 OR, while for others Gl = 4·0 OR.

Were it not for our knowledge that oxygen does not increase the rate of glycolysis in yeast we might have been inclined to suggest that in some way or other oxygen increases the efficiency of conversion of C to D. As this increase of glycolysis by oxygen is not merely a transitional effect, but leads to a permanent increase of adjusted rate, it is clear that the production of C as the substrate for glycolysis must be permanently increased by oxygen. We have therefore before us the possibility that oxygen has no direct effect on glycolysis at all, and that the rise of this functional activity in oxygen is merely the necessary outcome of the increased production of C. To this we shall return shortly.

The stages subsequent to glycolysis and their relation to oxygen. We have now to concern ourselves with the fate of the D group of substances which are the products of glycolysis and the reactants of the final stages of respiration. Our schema provides for their conversion either to NR, OR or OA. If we start our survey of oxygen effects with an apple in nitrogen we can regard the situation as simple: the whole of D that is being produced is converted to alcohol and CO_2 in the usual ratio, so that the consumption of D equals 3NR measured in carbon units. What then will happen when traces of oxygen are admitted by some low oxygen concentration outside the apple? We have already satisfied ourselves that this oxygen will not lower the rate of production of D, but nevertheless we find as a fact that the production of CO_2 falls off markedly, carrying with it, we assume, a similar drop in the production of alcohol. Now the substances of the D group are not autoxidizable so we naturally assume that the specific catalysts of oxidation are now able to start activity and compete with NR for D, oxidizing these substances to the extent that the oxygen supply can permit. We picture this oxidation as a reaction of high oxygen affinity and irreversible so that the consumption of the available oxygen is complete. As a result there is produced a certain amount of OR and OA. The totality of CO_2 escaping will now be partly derived from NR and partly from OR, so that we have use for the expression $TR = OR + NR$ (see Paper II, p. 41).

With the right external low concentration of oxygen it can be arranged that, say, half D is converted to NR + alcohol and half to OA + OR. With a still higher external oxygen concentration a point can be reached such that just enough oxygen enters the cells to convert the whole production of D to OA + OR, and there is now no longer any NR production. This marks a definite physiological state, and one experimental task in Paper V will be to locate the external concentration of oxygen that, for a given tissue, coincides with this 'extinction point' of NR. One physiological index of it, in a living apple, may be that at this point there is a minimum production of CO_2. Incidentally it has the significance that at this value there is presumably no longer any accumulation of alcohol in the tissues. There will be further consideration of extinction points in the following papers on oxygen. According to this view a higher concentration of oxygen, such as air provides, cannot carry out any more active oxidation of D, for by definition D is wholly oxidized to OA and OR by quite low supplies of oxygen, such as obtain at the extinction point, usually located in the region of 5% external O_2. In accordance with this we put forward the view that in any moderate concentration of oxygen the whole of the CO_2 is to be labelled OR, without any NR component. We have next to consider the magnitude of the OA production that is associated with this OR.

The magnitude of oxidative anabolism in respiration. We propose to treat OA as a substance, but it must be made clear that as far as evidence goes it is only a magnitude. In the section on glycolysis it was shown that by the application of a particular canon we claim to be able to measure glycolysis in each adjusted state of respiration in oxygen mixtures. Obviously we can measure OR in the same state, and OA is only the amount of carbon by which produced CO_2 falls short of carbon glycolysis. The amount is large, and as far as we have yet experimented we find that, for a given physiological state and class of apple, OA bears a constant ratio to OR. The extremes of this ratio observed for different classes of apples range from $OA = 3 \cdot 5\ OR$ to $OA = 3 \cdot 0\ OR$. With a given state of the apple this ratio seems to be independent of the magnitude of glycolysis and of the concentration of oxygen. So we adopt the view, provisionally, that the two substances are colligate parts of the products of one catalytic system and not the result of two independent catalysts working on D as substrate.

The next question to be faced is what becomes of OA thus continually being produced in oxygen mixtures, for there is no evidence of any down-grade product accumulating in the tissues. The symbol we have given to it is based on the assumption that it is anabolized back into the carbohydrate region of the system antecedent to glycolysis, so that possibly the same carbon atom may circulate round through the glycolytic machinery several times before it chances to be thrown out finally as CO_2. It might be that OA passed directly to the highest carbohydrates of group A, or to normal hexoses of group B, or, on the other hand, it might arrive at the bottom of the series in group C. As the amounts of A and B are very large, it would not make a significant difference to the working of the system should the return of OA to them be cut off for a period of hours. But should its destination be C, which can only be present in small amount, then the arrival or non-arrival of the OA contribution should make its mark on the concentration of C and therefore on the rate of glycolysis, which is a measurable function. It must be emphasized here that though our hypothetical substance OA is regarded as an outcome of the oxidation process OR, there is no reason to regard it as being necessarily an *oxidized* derivative of D. It might indeed be a reduced derivative, colligate with oxidized products $CO_2 + H_2O$. If that were so the catalytic process would be an oxido-reductive process and the reduced component be carried back to carbohydrate. A study of the actual oxygen consumption in all critical stages of apple respiration, which has recently been started, may help to throw some light on this uncertainty.

Finally, in this connexion, we have to refer to the existence in apples of a fair amount of a metabolite which is not a carbohydrate, namely, malic acid. If malic acid is to be identified with our OA it would be necessary to hold that this substance is always being rapidly produced and consumed, since it does not accumulate in starving apples, but diminishes slowly. The amount present at any one time would be the difference between the rate of production from D and the rate of consumption by the combined possible fates of further oxidation to OR and anabolism to A, B or C of the carbohydrate group. The amount of malic acid in the tissues can be directly determined, but its variations under change of oxygen concentration have not yet been established. There is ·a possibility of getting a further contribution to exact knowledge in this direction.

The stages antecedent to glycolysis and their relation to oxygen. The essential point for examination here is: does oxygen increase the production of C and, if so, through what mechanism is this brought about? The tendency of all the previous sections has been to make the concentration of C the real nodal point in the functioning of the catalytic system which we have schematized. An analysis of the rates of production and consumption of C is complicated by the possibility that no less than four reactions may be involved here. These are set out in schematic form on p. 64.

Two of them are founded on the expectation that the B − C relation must be a reversible one. A third is the irreversible glycolytic conversion of C to D, while the production of C from OA provides a probable fourth reaction. In nitrogen this set becomes simplified to three, for D goes wholly to NR, and no OA is produced. This power of cutting out OA by nitrogen is one of the experimental fields to which we can look for evidence as to whether OA should be held to pass to C. Any excessive production of C above a balanced state would lead to increased transport in the direction C→B, so that OA would find its way to B ultimately, and perhaps even to A. This distinction between the effects of OA and B as sources of C has been given some consideration, but as the specific velocities of the reactions concerned are unknown we must postpone the inquiry till the initial stages of transitions have been worked out for successive short periods, instead of the 3 hr. periods employed in the present investigation.

By subjecting an apple to pure oxygen instead of air we have established that glycolysis together with OR and OA are collectively increased 1·4 times, all keeping their relative proportions of $4·5 = 1 + 3·5$. To maintain this increased adjusted rate not only must C have a higher concentration but the production of C from A→B⇌C must be increased in the ratio of 1·4. Whether OA passes back

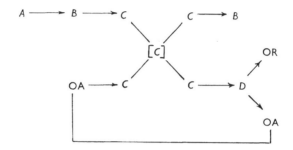

directly to B or at first appears as C does not affect the *adjusted* rates we are concerned with, though it might affect the precise form of the transition from one adjusted state to the other. We conclude from the above observations that the different oxygen concentrations in the cell bring about different rates of activity of the A − B⇌C reversible mechanism. Theoretically the effect might be attributed to depression of C→B or acceleration of B→C.

This effect may not involve the continuous chemical consumption of oxygen, and it might be attributed to activation of a catalyst system or to lowering of organization resistance. The change to the new rate of activity after alteration of external oxygen is extremely slow, lasting some 45 hr., during which time the OR output is observed to creep up or down to the new level in the form of an equilibration curve, so that at first it approaches the new position fairly fast and then slower and slower (halving time about 7·5 hr.). Such a slow rate of adjustment is characteristic of altered carbohydrate balance relations rather than of a direct chemical effect of altered rate of oxidation.

The implication, from all this, that the production of C from A and B is a function of the oxygen concentration, requires one marked qualification as regards very low oxygen supply. Without this qualification it might be falsely assumed that in complete absence of oxygen there would be no production of C and B, and therefore no glycolysis. These processes, however, proceed at a considerable rate in nitrogen; but this rate does not appear to be increased by very low supplies of

oxygen, but remains about the same minimal rate till the oxygen supply determining the 'extinction point' of NR is reached, say, 5% external O_2. It is
conceivable that up to that point, till NR ceases to be able to exist, the affinity for
oxygen is so great that practically zero partial pressure of oxygen is maintained
in the cells, so that only from the extinction point onwards is there sufficient free
oxygen to begin to cause that increase of C production from B which we have
just attributed to it. The facts on which these considerations are based will be
found in the later papers which deal with low oxygen concentrations. The
interpretation of this absence of effect with low oxygen may, however, turn out
to be metabolic.

We suggest that every increase of supply of *free* oxygen to the cell which raises
the partial pressure of oxygen above zero increases the rate of action of the pre-
glycolytic carbohydrate stage in the direction of increasing the production of the
substrate C for glycolysis. An inevitable consequence of this is that glycolysis
increases and more D is produced.

When we pass from higher oxygen to lower, then the glycolysis rate undergoes
a corresponding diminution which is seen in progress, in its simplest form, in the
passage from air to 5% O_2. When we pass from air to 0% O_2 the same lowering
of glycolysis takes place, but the evidence for it is obscured by the simultaneous
big change in CO_2 production, due to the effect of lack of oxygen on the post-
glycolytic oxidation stages.

In order to make a clear analysis of the complex changes that are at work
during transitions into or out of nitrogen, a full section will now be devoted to
a formal analysis of the varieties of transitional phenomena that our records
present.

3. The Form of Transitions with Change of Oxygen Supply

One of the outstanding characteristics of this investigation of respiration is the
long duration of the individual records. We obtain records of CO_2 production
lasting many days, and these show that after a change of oxygen condition the
respiration presently settles down to give a steady drift of values which is sometimes level but usually presents a gently declining curve. These states we speak of
as 'adjusted states', and in them all the stages of production and consumption
throughout the schema are proceeding at a practically uniform rate. When the
oxygen supply is suddenly changed we get phases showing a marked and sometimes a violent change of CO_2 production which phases presently come to an end
in new adjusted states. These phases we speak of as 'transitional', and the forms
of the transitional records call for a special survey, as it is from them that we have
derived suggestions for many of the points elaborated in this exposition.

There appeared, on our first inspection, to be two types of transition, and it was
a long time before we could provide an interpretation of their common features
and their distinctive features. We may label them as the *simple* type and the
complete type. The former manifested itself when a change was in progress from
one intensity of OR to another intensity of OR; while the latter characterized

changes which involved NR, such as change into nitrogen or out of nitrogen. There are many cases of the complex type in the data of the present paper; most of the simple transitions come up for treatment in the next paper, but two occur among the present data. These appear in fig. 1 as the first transitions of the two records. Apple XXIII goes through one on the change from air to pure oxygen

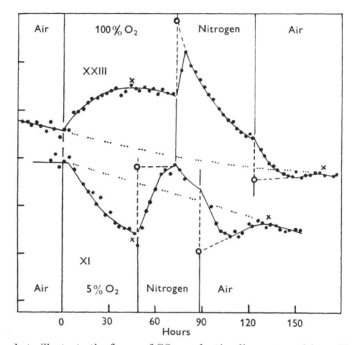

Fig. 1. Records to illustrate the forms of CO_2 production line at transitions. The single 3 hr. readings are represented by heavy dots, and these are connected up by a smoothed median line to bring out the form. The divisions on the ordinate axis are 4 mg. CO_2 apart; the actual values are not given here but will be found in the full records in Appendix I. The course of ALR is represented by groups of three small dots. The time period of 45 hr. after the beginning of certain transitions is marked by a cross close to the record. Each record presents three transitions, the first on each does not involve NR and therefore exhibits the *simple* type. The other two exhibit the *complex* type in which there is, in addition to the slow OR transition, a comparatively sudden change of post-glycolytic oxidation due to sudden appearance or suppression of NR with its higher production of CO_2. This sudden alteration to a new CO_2 level is indicated by the vertical broken line leading upwards or downwards to the circle which represents the initial value of the new state. These initial values are masked by the physical lag in CO_2 escape so that the extremely high or low early values cannot be observed. These are represented by the broken line after the initial point which presently joins on to the track of actually observed values. Record XXIII is typical in the form of its transition into or out of nitrogen, resembling the cases in fig. 2, but XI shows special features in its transition from 5 % O_2 to nitrogen which are discussed on p. 76.

and apple XI on the change from air to 5 % O_2. In XXIII there is an increase in OR from the air rate to the higher adjusted rate for oxygen, which is 1·4 times the rate for air; in the latter a decrease to adjusted rate for 5 % O_2 which lies at 0·73 the rate in air.

In the case of XXIII there is some clue to when the rising transition is over, because as soon as the adjusted rate is attained the record begins to decline. The transition is thus seen to be very slow, lasting at least 45 hr. It is hard to believe that it would take 45 hr. for a pure oxidation rate to adjust itself after an increase of oxygen concentration, and this is one reason why we attribute this observed change to carbohydrate metabolism. The form of the rising transition is exactly that of an approach to a reversible equilibrium, rising at first steeply and then slower and slower till the equilibrium state is reached. We picture the opposed processes to be increased production of C from A–B by oxygen activation, working against increased consumption of C by glycolysis, which rises with each rise of concentration until the two become adjusted to equality again.

In the case of apple XI, the lower record in fig. 1, the transition is downward from the air-line to the lower adjusted line for 5%. The form of the transition is exactly the same as that for XXIII, only the movement is all in the opposite direction. Though the end of the downward transition to a lower falling line is not so clearly located it is fairly well determined that this also lasts at least 45 hr. In confirmation of these forms of the simple type of transition we have observed the return from pure oxygen to air as proceeding just like the passage from air to $5\% \ O_2$, and the return from $5\% \ O_2$ to air like the change from air to $100\% \ O_2$.

When we observe transitions from air to nitrogen or the reverse we meet the complex form (see the first transitions of XXIa and XXIb in fig. 2). Though the change from air to nitrogen involves a lowering of the rate of production of C from A–B just as air to $5\% \ O_2$ does, yet the first effect on the CO_2 record is a rapid rise which comes to an end in about 9 hr. and is then followed by a long-continued fall in a curve of decreasing steepness. Every case of nitrogen given to an apple of class A or B provides an example of this. When the nitrogen is discontinued and the apple returns to air the record, instead of returning smoothly to the air-line, sweeps below it in 7–10 hr. or so, and then slowly climbs back to the air-line, as the third transitions in fig. 1 and the second in fig. 2. We have now satisfied ourselves that these forms represent the additive product of no less than three separate transitions, each with a different timing, and it is to this that they owe their complex form. Let us enumerate these three changes.

It is clear that the change from nitrogen to air should produce the same transitional type of pre-glycolytic increase of production of C from A–B as we have just fully described for the simple transition from air to oxygen. To this must be added a second transition due to the sudden entry of oxygen which proceeds at once to divert all D from its previous fate as NR to its new fate as OA + OR. As in this class of apple, D, when undergoing NR, gives one-third of its carbon as CO_2, and when diverted to OA + OR gives only $1/4\cdot5$ of its carbon as CO_2, there must be a sudden drop in CO_2 production within the tissues to two-thirds of its previous rate. This process if it stood alone would have as its transition an almost sudden drop of CO_2 production. The combined transition of these two would take the form of a sudden drop of CO_2, due to the post-glycolytic oxidation change, down to a value which was two-thirds the last NR value before the admission of oxygen, and from this low point there would develop the slow rise of CO_2 lasting 45 hr. due to the simple transition of increasing C in the pre-glycolytic stage. This combined form is drawn in figs. 1 and 2 as a broken line dropping vertically to a circle locating the new theoretical initial and thence rising slowly towards the new adjustment.

The third transitional effect which comes in to mask the combination of two just described is a purely physical affair, due to the long time it takes to equilibrate the CO_2 content of an apple with that of its environment by diffusion across the surface, which offers a considerable resistance. If the internal production of CO_2 were cut off suddenly and completely, the store of CO_2 in the watery tissues would go on escaping giving a geometrically decreasing curve of CO_2 escape to the exterior, and it would be very many hours before this had all escaped to the air current. During all this time there would be an appearance of decreasing CO_2

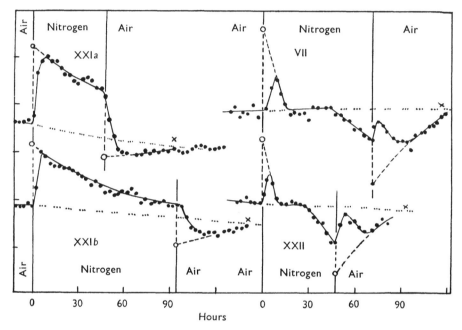

Fig. 2. General description as in fig. 1. Records XXI*a* and *b* exhibit typical complex transitions. The transition air to nitrogen for VII and XXII is also of this type, but the transition from nitrogen to air contains, here, an 'after-effect of NR' giving a small temporary extra production of CO_2. The last two cases, being class C apples, are described later (see p. 77).

production by the apple. The same form of CO_2 escape, though in a rising sequence, would accompany any instantaneous increase in actual metabolic production, making the rise appear slower than reality, and thus there is always an external distortion of the true form of the change of rate of internal CO_2 production. It follows that the sudden drop in the CO_2 production rate, due to the change from nitrogen to air, reveals itself only as a declining rate of escape of moderate duration, taking at least 10 hr. and often more in an imporous apple till the new lower rate of actual production is attained.

Combining this third transition which is the physical CO_2 transition with the sudden oxygen transition lowering the metabolic CO_2 production from D and with, also, the glycolytic transition, i.e. the slow rise of glycolysis due to increased production of C, we can build up exactly the observed form of transition in CO_2 escape from the apple which has been recorded many times at the change from nitrogen to air; see records XXIa, XXIb, XI and XXIII in figs. 1 and 2.

These triple complex transitions are not quite the end of the complication, for in several living tissues other than apples that we have studied, the most marked effect of air after nitrogen is a very large sudden production of CO_2, due to the oxidation of some accumulated product of NR metabolism. This we speak of as the 'after-effect' of NR. In most metabolic states of apples there is no trace of such an after-effect, but apples of class C show a very small after-effect (see records VII and XXII in fig. 2), and so also does the very late apple XXV of class B. In such cases there is a temporary rise of CO_2 production at the transition, maximating almost at once and subsiding in about 24 hr., just in time to reveal the last stage of the triple complex transition that we have described in detail. The end of these various transitions is that at last the CO_2 production works its way up to the adjusted air-line value ALR, and then drifts slowly down it in continuation of the original direction of the line before nitrogen was given. The nitrogen experience is at last, after a couple of days, a thing of the past which has left no permanent effect.

The contrary transition, from air to nitrogen, to which we now turn should clearly be built up as a complex whole from the three constituent transitions, all working in the opposite way. And this is what it proves to be. In the two figures the broken-line course and the circle set out the proposed interpretation of these transitions. There is the quick oxidation transition, which at the cutting out of oxygen by nitrogen should increase the CO_2 production from D suddenly to 1·5 times its OR value. But this high initial value is not actually attained on account of the physical CO_2 transition, which makes a lag of about 10 hr. before the rapidly rising CO_2 escape gets to its highest value. All this time the production of C from A–B is slowly declining towards the low rate characteristic of nitrogen. If there were certainty as to the direction and ultimate course of the 'adjusted nitrogen line' towards which the transitional complex is working, then we could determine when the whole transition should be judged to be over. Certainly the falling NR series is still slackening off at hour 45, and even at hour 100 it is not certain we have a true adjusted rate. It may be that in time there is a toxic effect of accumulated alcohol and that no really adjusted rate can be maintained. Therefore at present, till longer records in nitrogen have been obtained, we propose to speak of the nitrogen values after 70–100 hr. or so as giving a 'semi-adjusted' nitrogen line.

So far we have only dealt with the nitrogen transitions of apples of classes A and B. It will, however, not have been forgotten that Class C presents us with nitrogen effects that at first seemed to be quite different. In reality these also can be satisfactorily interpreted on exactly the same analytic lines. Cases are presented in fig. 2, records VII and XXII, but we shall not bring them to account here, as a full treatment of them will be found in connexion with the study of glycolysis in apples of class C at the end of the next section of this paper (see p. 77).

4. The Transitional CO_2 Ratios of the Individual Records and the Construction of their Glycolysis Lines

Having provided an interpretation of all the features of the CO_2 records at transitions from one gas mixture to another, distinguishing the effects of oxygen on pre-glycolytic carbohydrate phases from the effects on post-glycolytic oxidation phases, we can now enter upon a further development. This takes up the problem of evaluating the rate of glycolysis in the different individual records, on the basis of their individual transitional CO_2 ratios.

The important new line that we claim to be able to add to these records is the *glycolysis line*, which sets out the rate of the glycolytic conversion of C to D throughout the whole course of the record, whatever may be the gas mixture surrounding the apple. The theoretical significance of glycolysis rate has already been elaborated. We regard it as a measure of the concentration of its substrate C, which is the outcome of the production activity of the carbohydrate metabolism chain A–B–C. The form of this line therefore reveals not only the general effect of starvation drift in lowering activity, but also the special effects of oxygen, air and nitrogen upon the rate of glycolysis. By concentrating attention on glycolysis we get away from the distractions of the different CO_2 production ratios of NR and OR and contemplate the influence of the various oxygen mixtures upon metabolism apart from oxidation. From this point of view CO_2 production serves merely as an index of the rate of glycolysis, but the index value has to be weighted differently according to whether it is CO_2 of NR or CO_2 of OR. When this is achieved it is brought out that the form of the glycolysis line is much simpler than that of the CO_2 production line. This is because it succeeds in presenting us with that greatest common measure of respiration, be it NR or OR, which we set out to seek in our analysis.

In proceeding from the CO_2 line to the glycolysis line we were working our way backward, by analytical treatment, from observed results to hidden cause. If, having attained a comprehension of the glycolysis line, we now turn round to contemplate the devious course of the CO_2 line we see how its distortions reveal the magnitude of effects of post-glycolytic phases in the catabolism, phases in which, according to the oxygen supply, oxidation of the products of glycolysis D, may be either complete, partial or entirely absent.

For each analysed record, we shall have to give the quantitative relations upon which we have based our construction of the glycolysis line. The general nature of these numerical relations may be set out here so that only the values need be stated for each record. For the evaluation of glycolysis while in nitrogen we take it that three times the values of NR gives us an accurate measure, provided the rate of CO_2 escape is properly adjusted physically to the rate of NR CO_2 production. For the first 8 hr. or so after entry into nitrogen we have a transitional phase, and the physical CO_2 lag conceals the full rate of NR. The initial NR, which is a value of vital analytic importance, is obtained only by extrapolation to zero hour. Glycolysis at the moment of transition is $3 \times$ initial NR; this product must measure also the glycolysis rate in the oxygen mixture before nitrogen. From the numerical relation of the final OR value to the initial NR we get the ratio which enables us to determine the magnitude of glycolysis in air for that particular case.

If $\dfrac{\text{initial NR}}{\text{final OR}} = 1\cdot 50$, and $\text{Gl} = 3\,\text{NR}$, then $\text{Gl} = 4\cdot 5\,\text{OR}$.

On the return from nitrogen to air we get another opportunity of checking the value of this ratio with lower absolute values of NR and OR, by the ratio final NR/initial OR. In this situation, it is the precise initial value of OR which is masked by the CO_2 physical lag, now working in the opposite direction, and we again depend upon extrapolation to zero hour of the rising OR series.

Table 1. *Table of transitional values and ratios*

Apple	$\dfrac{\text{Initial NR}}{\text{Final OR}}$	Hours in nitrogen	$\dfrac{\text{Final NR}}{\text{Initial OR}}$	Ratio	Class
XIX	$\dfrac{22\cdot 0}{14\cdot 2}$	70	$\dfrac{17\cdot 0}{11\cdot 0}$	1·55	A
XXIa	$\dfrac{18\cdot 85}{12\cdot 4}$	48	$\dfrac{14\cdot 9}{9\cdot 8}$	1·52	A
XXIb	$\dfrac{14\cdot 65}{9\cdot 7}$	96	$\dfrac{9\cdot 35}{6\cdot 2}$	1·51	A
XI	$\dfrac{[16\cdot 40]}{10\cdot 9}$	40	$\dfrac{14\cdot 5}{9\cdot 7}$	1·50	A
XXIII	$\dfrac{24\cdot 0}{18\cdot 5}$	50	$\dfrac{15\cdot 0}{11\cdot 5}$	1·30	B
XXV	$\dfrac{16\cdot 6}{12\cdot 5}$	50	$\dfrac{11\cdot 5}{8\cdot 6}$	1·33	B
VII	$\dfrac{20\cdot 4}{13\cdot 6}$	70	$\dfrac{11\cdot 1}{7\cdot 4}$	1·50	C
XXII	$\dfrac{21\cdot 3}{16\cdot 1}$	50	$\dfrac{12\cdot 7}{9\cdot 6}$	1·32	C
XXIV	$\dfrac{27\cdot 6}{20\cdot 8}$	112	Killed	1·33	C

Initial NR values obtained by extrapolation. Final OR and final NR values observed. Initial OR values calculated by the initial transitional ratio and fitness judged graphically.

In table 1 we have set out the data for all the cases of nitrogen respiration examined in the previous paper. The second column contains the NR and OR values adopted at the change into nitrogen, while column 5 gives the resulting transitional ratios. The ratios of Gl to OR will be threefold these ratios. Column 4 contains, as denominators, the values of initial OR on passing out of nitrogen, which are indicated by the application of the adopted ratios.

The value of the OA component is evaluated from the same data being in terms of carbon units, Gl less OR in any oxygen mixture; and in nitrogen Gl less $3\,\text{NR}$, this latter being, by our postulates, zero. Finally, we can state the ratio OA/OR for each case.

Apples of classes A and B. We may begin our survey with the apples of the classes A and B which have already passed into their late starvation stage, and consider first those only involving air and nitrogen.

Records XXI*a* and *b* (see fig. 3). On the falling CO_2 production line in air we have as final OR value 12·4. With entry into nitrogen we get a steep rise in CO_2 to reach the falling NR line, which line by extrapolation gives initial NR = 18·85. The transitional ratio for air to nitrogen is therefore established as 1·52. At the end of nitrogen NR has fallen to 14·9, which if the same ratio holds would indicate 9·8 as the initial OR value. This value joins up well enough with the observed

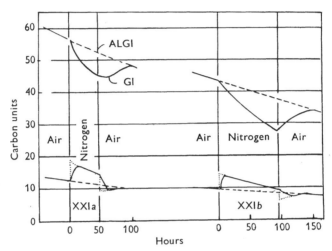

Fig. 3. This and the next two figures illustrate the construction of the glycolysis lines from the CO_2 records. Here the ordinate values are carbon units so as to give a possible common measure for CO_2 output and glycolysis. The unit is the amount of carbon in 1 mg. of CO_2, i.e. 0·27 mg. carbon. The ordinate numerals are therefore the same as those for CO_2 in the general records. The lower part of each figure gives the rate of CO_2 production treated as in figs. 1 and 2, while the upper part gives the evaluated glycolysis lines, Gl being a continuous line for the actual glycolysis and ALGl a broken line for the 'air glycolysis line', which is the counterpart of the ALR for respiration.

sequence of rising OR values, so we regard the ratio as sufficiently established and are prepared to draw the glycolysis line at $3 \times 1·52$ OR for all times when the apple is in air. When in nitrogen Gl = 3 NR, so we can carry the line right through the record. This glycolysis line, Gl, appears in the upper part of the figure as a continuous line, in connexion with the broken line representing the glycolysis air-line ALGl.

Turning to the second nitrogen experience of the same apple, XXI*b*, some 130 hr. later we find final OR = 9·7, initial NR = 14·65, giving thus a CO_2 ratio for air to nitrogen of 1·51. At the end of 100 hr. in nitrogen NR has fallen to 9·35, which on applying the same ratio indicates an initial OR of 6·2 which takes its place perfectly as the start of the rising OR values making back to the air-line. We have then four evaluations of this ratio for one apple at remote periods of time, namely, at hours 48, 95, 234 and 334 from the time of removal from store at 2·5° C. to the experimental chamber at 22° C. We consider it established by this record that this fundamental ratio is practically constant for a given apple and not subject to serious time drift. The ratio once ascertained for a given apple we can then use OR as a satisfactory index of glycolysis and its drift.

We may recall that this glycolysis line is not really a respiration line at all; it is a line of drift of carbohydrate catabolism. Its drifting reveals that nitrogen depresses it—by retarding the rate of production of the substrate C from A–B, according to our interpretation of the situation. When air is readmitted the rate of this production rises, and goes on rising till the production and consumption are balanced once more along the old line of starvation drift ALGl. This recovery of glycolysis rate takes about 45–50 hr. in both cases and exhibits a simple transition. But when we look from the glycolysis transition to the identical transitional state on the CO_2 record in the lower part of the figure we find a triple complex transition, in which the drop of CO_2 production from Gl/3 in nitrogen to Gl/4·5 in air is very quick; while the transition of physical escape of CO_2, which is of moderate duration, also comes in to mask the metabolic events in the manner set out in the previous section.

With regard to the transition into nitrogen from air the CO_2 record shows an inverse triple complex transition, but from it there emerges the simple transition of the glycolysis line. We are, however, not able to state when this transition comes to an end, as the lines curve down still after 45 hr. and passes over into what we describe as a 'semi-adjusted' course.

Let us pause for a moment to consider what the curve of *carbon loss* for these cases would be, though we have not drawn this curve in the figures. Its course presents no subtleties, since in air it is identical with OR, being just the carbon contained in the escaping CO_2, while in nitrogen it is identical with Gl, the whole carbon of D being lost, either as alcohol or CO_2. At the transition to nitrogen the carbon loss shoots up from being identical with the lower record to become identical with the upper, where it remains until air is readmitted; after which it drops back to the lower again. Such a record would serve to bring out the enormous increase of loss that at once takes place in nitrogen, but it is not yet clear whether this great loss is a factor in causing the continuous rapid decline of glycolysis in nitrogen. According to the views already elaborated, this big loss comes about by nitrogen cutting off the formation of OA at the same time as OR, so that there is nothing produced in nitrogen which can be built back by anabolism to the carbohydrate group. We must therefore recognize the big part that oxygen plays in restraining carbon loss, when the state of things in air is compared with the state revealed in nitrogen.

The measure of the large amount of OA that is held to be formed in air is easily estimated by eye without an independent line, since it is always Gl less OR. In these records OA is 3·5 times OR.

Record XIX associates itself very closely with these two, so much so that it might almost be an earlier nitrogen experience of the same apple. It is, however, only a short record, with no complete recovery to air after nitrogen, so that it will not be figured here. Also it must be borne in mind that its NR course has been very much smoothed (see Paper II, fig. 3). For XIX the last value in air before nitrogen was $OR = 14\cdot2$, while the extrapolated initial of the smoothed curve was $NR = 22\cdot0$, giving a transition ratio of $1\cdot55$. After 70 hr. in nitrogen the last NR value is $17\cdot0$, so that by the same ratio the initial OR after nitrogen should be $11\cdot0$, which is not out of harmony with the imperfect record, cut short by failure of temperature control. We conclude that glycolysis is $3 \times 1\cdot55\,OR$ in this case.

Record XXV is that of an apple which has been assigned in Paper I of this series to class B, on the evidence of the pitch and drift of its air-line. It provides a single nitrogen experience, 50 hr. long (see fig. 5, p. 77). The final OR value before nitrogen was $12\cdot5$, and the initial NR $16\cdot6$, which gives us a ratio of $1\cdot33$. Later, the final NR was $11\cdot5$, indicating, by the $1\cdot33$ ratio, an initial OR of $8\cdot6$. The transition from air to nitrogen is a typical triple complex one, but the transition from nitrogen to air is an example of the quadruple transition that we described in the section on transitions, for here there is an additional feature of a small but definite after-effect (see p. 69). This hinders us from getting clear support for the suggested transitional ratio at the end of nitrogen, but $OR = 8\cdot6$ serves well enough as an initial value. The glycolysis line for this B apple is constructed from the values $Gl = 3 \times 1\cdot33\,OR$ and $Gl = 3\,NR$. The decline of glycolysis in nitrogen is seen to be much steeper in this record. Here OA, calculated by Gl less OR, has the value of $3\cdot0\,OR$. We suggest that the $1\cdot33$ ratio is possibly characteristic of class B as opposed to the group of ratios about $1\cdot5$ found for class A.

Record XXIII is also for an apple assigned to class B. It presents a new experimental feature in that the apple was for a long time in pure oxygen before experiencing nitrogen (see fig. 4). On passing from air to oxygen the CO_2 rises smoothly, at first steeply and then slower, till a crest is reached at 45 hr., at an OR value of $19\cdot0$ which is just $1\cdot4$ times the ALR value of $13\cdot8$ at that time. Then for 30 hr. the OR declines maintaining the same ratio to ALR constantly. Here then we find the adjusted pure oxygen line as a constant ratio to the air-line. This rise of CO_2 production produced by increase of oxygen has quite a different form of transition from the rise of CO_2 of about the same magnitude produced by nitrogen. It is of fundamental importance to ascertain whether glycolysis has gone up with the rise of OR, or whether this increased OR is due merely to heightened oxidation of D, the glycolysis remaining the same as in air. The nitrogen experience which follows settled this point quite definitely, for extrapolation of the NR series gives initial $NR = 24\cdot0$, the highest value ever recorded; while the last OR value was $18\cdot5$, indicating a transitional ratio of $1\cdot30$, which concords with the ratio of the other B apple XXV.

We note that at the time of transition the value on the ALR line is $13\cdot2$, which gives a ratio for NR/ALR of $1\cdot82$. It is this ratio that attention was focused on in Part II of Paper II, following the usual tradition; but now we learn that when the apple is in oxygen glycolysis is not the same as in air, and the nitrogen/air ratio loses any individual significance, being only the product of the OR/ALR ratio of $1\cdot40$ and the NR/OR ratio of $1\cdot30$.

Clearly glycolysis at the transition is $3 \times 24\cdot0$ and has risen *pari passu* with the rise of OR, maintaining the appropriate $1\cdot30$ ratio. We require yet confirmation of the $1\cdot30$ ratio as specific for this apple, and this we get at the transition from nitrogen back to air. The final NR value is 15·0, and the form of the rising OR values confirms the initial OR value of 11·5, which the 1·30 ratio would indicate if it held throughout this apple record. We conclude then that, however high the value of OR may be forced, glycolysis rises in the same ratio continuing to be evaluated by Gl = 3 × 1·30 OR.

In thus speaking of OR as if it were a determiner of glycolysis we must realize that we are inverting the real sequence of causation, for we regard OR merely as an index and not a cause of glycolysis. Also glycolysis in its turn is taken to be only an index of the production of C. If our views were absolutely established our statement would take the form that pure oxygen has increased the production

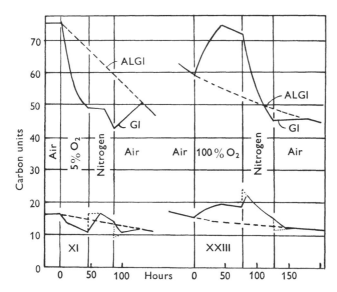

Fig. 4. General description as in fig. 3. Records of XXIII and XI each begin with a simple transition followed by two of the complex type.

rate of C to 1·4 the value in air, and that the rates of glycolysis and OR rise consequently in the same proportion, OR continuing to be one quarter $\left(\dfrac{1}{3} \times \dfrac{1}{1·30}\right)$ of glycolysis all along. Hence OA = 3·0 OR as contrasted with OA = 3·5 OR for class A.

All glycolysis transitions are, by definition, simple in form and slow in progress, but with this apple on passing into pure oxygen from air we have met for the first time an OR transition which is also simple and slow. This is confirmatory of the view that the complex form of transition is to be attributed to the intervention of NR, causing a shift of the post-glycolytic relations.

This record contains the most striking variations of glycolysis rate that we have yet met, and we note how steeply the rate falls off in nitrogen after the height to which it has been pushed in pure oxygen.

By giving oxygen to raise glycolysis, and then nitrogen to lower it, it is possible to meet a moment of time in nitrogen at which the glycolysis rate, being still unadjusted, just happens in passing to attain the value appropriate to air at that time. This is located in the figure at the moment when the falling glycolysis just cuts across the line that glycolysis in air would have followed. According to our present views, return to air at that moment should arrest Gl at that value and maintain it there. The observed CO_2 would be expected to drop from $NR = 1.30\,OR$ to OR and stay there without dipping below ALR and creeping up again. This would be an interesting region for future experimentation; for here a change in oxidation might be examined apart from any change in glycolysis rate.

There is still one record left to be considered, apple XI, also presented in fig. 4. This differs from all the others in the fact that it is an apple in the early stage of storage and not in the late starvation stage as the other five. In the first paper the characteristics of the early phase are fully discussed, but as XI stands alone in regard to the nitrogen problems, it is not easy to be sure of the interpretation of its behaviour, even though it has been identified as belonging to class A. Another feature in which it has no parallel is that it was taken into nitrogen direct from $5\% \ O_2$ which is a concentration close to the extinction point of NR. Its record begins in air, but quite soon the oxygen concentration was suddenly lowered to 5%. We see the CO_2 production falls steadily as we should expect. This diminishing OR falls off at first fast and then slower, reaching finally the adjusted state for 5%, which lies at about 0.73 of the ALR values in air. The transition to this lower adjustment takes about 45 hr., and though reversed in direction is of simple form quite similar to the rise of OR in XXIII, indicating that this case like the other is due to equilibration of the carbohydrate production of C. The last OR value in 5% before nitrogen is 10.9; what initial nitrogen value is to be expected? As this is a class A apple the expectation might be $NR = 1.5 \times 10.9 = 16.35$.

There is, however, a special feature about the NR series in this case in that the CO_2 rise is exceptionally slow, taking about 24 hr. to reach its highest value, but the value when reached is exactly the value indicated above after which the NR values fall off. What course are we to give to the early curve extrapolated back for 24 hr. to zero hour? We may first seek support for the use of the 1.5 ratio by examining the transition *out* of nitrogen into air. Here the last NR value is 14.5 which, divided by 1.5, indicates 9.60 as the initial OR in air. This value joins up perfectly with the observed low OR values making a typical 45 hr. transition to the ALR line. Clearly the 1.5 ratio is appropriate to apple XI, and we incline more strongly to adopt the initial NR value of 16.35 as shown in the figure. The consequence of this is that the extrapolated NR series runs a level course for 24 hr. Such level values are characteristic of apples of class C, which long retain their youthful condition, so they may appear here also for the one early apple of class A. This case must be left for later investigation, for our present interpretation contains hypothetical elements for which we have no experimental support from the other records. A thorough-going application of the 1.5 ratio gives us $Gl = 3 \times 1.50\,OR$ as fixing the glycolysis line shown in the figure, covering not only the period in air but also the fall of glycolysis in 5% to the initial value in nitrogen. A curve for the rise of glycolysis from the end of the nitrogen experience back to the air glycolysis line is also drawn.

The nitrogen records of apples of class C. The nitrogen records of the three apples assigned to this special class were compared with those of classes A and B in the opening pages of Paper II, and at first sight the classes appeared to have no feature in common. It is perhaps significant that it was on the evidence of the course of the respiration in air that this class was set apart in the first paper of this series. That course is a result of carbohydrate metabolism and of the supply of substrate for respiration. Since we have now come to the conclusion that the oxygen-nitrogen behaviour of apples is primarily an outcome of the effects of these gases on the carbohydrate equilibration, which controls glycolysis, it is no longer surprising to find that a class which shows a special starvation behaviour in air should also show special features in nitrogen. Taking up the method of analysis that has been worked out for classes A and B we now find that it clarifies the

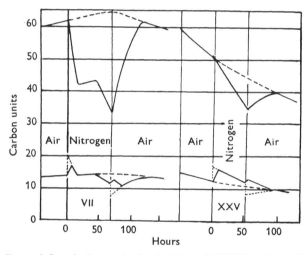

Fig. 5. General description as in fig. 3. Record XXV has been discussed earlier (see p. 74).

nitrogen behaviour of class C to such an extent that we can now regard this as a special case of the general formulation and no longer a type quite apart.

The three apples of this class that were subjected to nitrogen bore the numbers VII, XXII and XXIV, one being very early in the storage series and the other two very late. The transitional features of VII and XXII are to be found in fig. 2, and the glycolysis line of VII in fig. 5 provided with construction lines of the type we have already described. The special features of these apples that arrested us, on empirical survey, were that the CO_2 production in nitrogen is identical with that in air, for a considerable period of time, and that then, quite suddenly, NR drops away rapidly. Added to these two features was the appearance of a special transitional effect in that the CO_2 output rose at first for a few hours in nitrogen and then fell back to become again level with the OR values before nitrogen. This feature was at first put aside as some obscure disturbance of CO_2 production at the cutting off of air, but now that we have elucidated the components of the complex transitions of classes A and B we acquire the key to the significance of the form in C. It seems reasonable to interpret the initial form which gives a rise of CO_2 for some 8 hr. followed by a steep fall, as being really the expression of a high initial NR falling steeply away, distorted by physical lag in CO_2 escape from the apple. The magnitude of the initial NR was therefore to be arrived at by extrapolation of the observed falling slope back to the origin and a value obtained for comparison with the final OR. We could thus evaluate the ratios which gives a measure of the rate of glycolysis in air (see table 1, p. 71).

Record VII treated in this way gives us an initial value of NR = 20·4, while the final OR was 13·6. We get from these the same ratio of 1·50 that we got for class A apples. We conclude that here also glycolysis = $3 \times 1·5$ OR while in air. This ratio is supported fairly well by an examination of the ratio at the end of nitrogen where the final NR is 11·1, indicating initial OR of 7·4. The evidence cannot be decisive here because in this apple as in the other apple of class C there is superposed on the complex triple transition a small after-effect of exactly the type described for apple XXV.

On the sum of this evidence we may construct the glycolysis line of VII as equal to 4·5 OR in air and 3 NR in nitrogen. Clearly the form of this line in the upper part of fig. 5 is not fundamentally different in its gas relations from that of classes A and B, but it does present a special secondary difference in this, that instead of falling away in nitrogen along a smooth curve from the initial it first falls fast and then maintains a short level phase for a period of time before the fall is continued, rather suddenly and rather steeply. In our view this behaviour must be an expression of the factors governing the supply of the glycolysis substrate C. It is a difference of carbohydrate metabolism and not of respiratory oxidation. The relations of Gl to OA and OR, and of Gl to NR are the same as in class A. The characteristic thing is the form and rate of the depression of glycolysis on continuance in nitrogen. Whereas, with apples of class A, glycolysis in nitrogen is depressed in a smooth falling curve at such a rate that NR, which equals Gl/3, never gets as low as OR in the experimental period, here in class C we find that 15 hr. suffices to depress glycolysis so that Gl/3 is just equal to ALR, while after 45 hr. this value falls away fast to below ALR.

The rapid decline observed in these apples seems associated with a rapid power of recovery of glycolysis in air, for after nitrogen the original rate is built up again in about 45 hr. in spite of the very great previous depression.

This form of the depression of glycolysis which appears as exceptional among our apples, occurring only in association with those apples that can keep up a level respiration in air, has been found by our investigation of other plant tissues to be the typical form in leaves of cherry laurel. In these leaves, when starving, carbohydrate metabolism runs a different course from that of starving apples, as shown by their records of respiration in air, and in them nitrogen invariably produces exactly in every detail the form of effect here described for apple VII. The significance of this form will be taken up again in our analysis of the respiration of these leaves in a future paper.

The two other apples of this class must be briefly examined. Apple XXII agrees exactly with VII in type, but can maintain the level middle period in nitrogen only for a short time; the rectilinear fall sets in after 22 instead of 45 hr. Its transitional behaviour appears in fig. 2, but its glycolysis features agree so well with those of VII that no figure is presented. The extrapolation of the first falling NR values points to an initial NR value of 21·3, which gives a ratio of 1·32 to the last OR value of 13·6 before nitrogen. The glycolysis line would therefore be drawn as $Gl = 3 \times 1·32\,OR$. At the end of nitrogen we find final $NR = 12·7$, indicating by this ratio an initial OR of 9·6. This fits in well enough with the rising curve of OR values, reaching the air-line in about 45 hr.; but the situation is obscured, as in VII, by the presence of an after-effect.

Apple XXIV gives a fragmentary record which is therefore not figured except in the general records. It was not returned to air after nitrogen, as we wished to follow the nitrogen depression which here was associated with toxic effects. The early fall of NR is very steep and the extrapolation to the origin is not very securely located, but we accept 27·6 as the initial NR. This gives a ratio of 1·33 to the final OR of 20·8. The glycolysis line would therefore be constructed at $Gl = 3 \times 1·33\,OR$.

In record XXIV the level middle phase in nitrogen is so short on the NR record that its existence might have been ignored had it not been for the indications of the records of VII and XXII. The decreasing duration of this middle phase along the chronological series we should attribute to altered carbohydrate metabolism with metabolic drift. We may note that in apple XXIV the form of the NR progress has become practically a continuous curve, which brings it within the formal definition of class A, though very much steeper.

Surveying these ratios for the three class C apples we note that VII gives 1·50, while XXII gives 1·32 and XXIV gives 1·33. We find then both types of ratio in one class, but it must be mentioned that 7 months elapsed between VII and the two later examples. There is possibly a correlation of the lower ratio with late metabolic states.

5. Conclusion: the Outstanding Features of the Proposed Respiration Schema

In the first section of this paper we set out dogmatically a proposed schema of respiration reactions; in the second section we sketched our view of the functional working of this chain of reactions in relation to starvation and to variation of oxygen supply; in later sections we turned to the actual records, and demonstrated that the system provides a plausible interpretation of all the quantitative variations of CO_2 production that we had observed in these apples.

Now that we have worked through all these particular aspects of the matter we may conclude by surveying more generally the essential features of the new situation. The most fundamental departure is that attention is concentrated upon the rate of glycolysis, as much when the tissue is in air as when in nitrogen. Glycolysis is regarded as the common measure of respiration in all conditions. The production of CO_2 provides us with an index of the magnitude of glycolysis.

Another feature is the adoption of the view that normal hexoses (group B of the schema formulated on p. 56) are not the direct substrate for glycolysis or oxidation but that some specialized derivative of them is indicated for this function. This we represent by our group C, suggesting that the more active heterohexoses might prove a suitable representative for this position. The general reasons for interpolating a new reactant in the series are based on dynamical considerations. The way the apple metabolism responds to changes of oxygen supply suggests that the significant reactant is not a substance of high concentration which can undergo only slight alterations of amount, but rather one of low concentration which is subject to marked changes of concentration, production, and consumption within the range of the experimental changes we have imposed upon our apples.

It is clear that we can alter the glycolysis rate from a minimum of unity to a maximum of threefold. Postulating that the mass laws hold in this metabolism, at least a similar shift in the concentration of the substrate is required. We have no estimations of carbohydrates available for these apples, but general knowledge of sugar content gives no support to the view that the amounts of normal hexoses is sufficiently mobile. Exact knowledge in this field is being rapidly acquired by direct and continuous analytic studies of sugars in apples carried out by other workers in connexion with the Food Investigation Board; and from this survey it should be possible to make a final decision by combining carbohydrate analyses with direct respiration estimations.

In suggesting that the concentration of a substrate C, of which we have no knowledge in apples, is really the important aspect of all types of respiration, we have simply followed up the indications that we draw from our data. We have data from other tissues awaiting similar analysis, so that later we shall see where they in their turn appear to lead us. After that it may become necessary to take the present schema to pieces and reconstruct it, but at least we shall have consolidated a mass of relations to which any future system must conform.

Another new feature which arises out of the analysis of apple respiration and the concentration of attention on the rate of glycolysis is that in air a large amount of some substance, which we have labelled OA, is formed concomitantly with the $CO_2 + H_2O$ of OR. This conception, again, arises out of the facts presented to us and seems unavoidable, but as to the exact status and fate of OA we have not yet any decisive evidence. That it does not accumulate seems certain, and so we consider it as being built back into the stream of catabolites. It would be possible to hold that OA results from a catalytic oxidation process, which is independent of OR but has identical oxygen relations, or that OA is an antecedent of OR, so that in air the whole of glycolysis goes to OA first, while only part of it is oxidized on to CO_2. If the concentration of OA were low with a specific catalyst converting it to OR, then a constant relation between production of OA and OR might still be maintained through considerable variations of the rate of total glycolysis. Even within the range of variation of metabolic types presented by this one lot of apples we find evidence that the ratio of OA to OR may range from 3·0 to 3·5, when types are contrasted, though it seems to remain constant within a given type. With other plant tissues we may find such a variation of this ratio that the production of OA and of OR will cease to be regarded as two colligate aspects of one catalytic activity.

In spite of the uncertainties that surround the new components C and OA, introduced into our survey of respiration, we hold that one definite advance has been made by showing how glycolysis in oxygen mixtures can be evaluated, and a second by reaching the conclusion from such evaluations that glycolysis proceeds at a greater rate in air than in nitrogen, and is still further accelerated by further rise of oxygen concentration. It is here suggested that this acceleration of glycolysis by oxygen is not due to oxidation but to the acceleration of the rate of production of the substrate for glycolysis. Should it be established that the primary effect of varying oxygen supply in respiration lies in the control of carbohydrate equilibrium, then our biological outlook on this function will be considerably modified.

IV

FORMULATION OF THE EFFECTS OF LOW OXYGEN SUPPLY UPON CARBON DIOXIDE PRODUCTION IN APPLES*

CONTENTS

INTRODUCTION

1. *Experimentation and procedure*

This paper is a direct continuation of the three that were published in October 1928, entitled, 'Analytic Studies in Plant Respiration, I, II and III'.† It carries on further the study of the respiratory phenomena presented by apples under variation of the conditions of external oxygen supply. The earlier papers dealt with the effects of air and of nitrogen, while the present one takes up the behaviour of metabolism when supplied with intermediate low concentrations of oxygen. It must be understood that all the experimental data dealt with in the set of three papers, now presented as Papers IV, V and VI, form part, with papers I, II and III, of a single experimental sequence designed to explore the effect of varied supplies of oxygen upon respiratory metabolism as indicated by CO_2 production, in the tissues of a higher plant‡

The whole investigation was carried out at 22° C., and for each experiment only one individual apple was used. The detailed records of the respiration of the twenty-one individuals were brought together as an Appendix§ to the second paper of this series, Paper II, the individuals being throughout identified by the Roman numerals V–XXV. That Appendix can be referred to for full details of characterization, date and experimental treatment.

The general technique of experimentation and of accumulating data of CO_2 production is identical with that described in the first and second papers of the

---

* This paper has not been published previously. In the typescript the authors were given as P. Parija and F. F. Blackman. The following note in Blackman's writing was on the typescript: 'Read through and passed, 4 Oct. 37'.

† *Proc. Roy. Soc.* B, **103**, 412–523, 1928. As mentioned in the Preface these papers are republished as the first three papers in this book.

‡ For the general outlook that informed this study of respiration, see the opening paragraphs of Paper I.

§ The Appendix appears at the end of the book as Appendix I.

series. The experiments involving gas mixtures containing less oxygen than the 21 % in air were carried out some years ago before prepared low concentrations of oxygen were available in commerce, compressed in cylinders. These mixtures had therefore to be prepared for each experiment by adding nitrogen from cylinders to air in a graduated gas-holder, till the desired dilution of O$_2$ was attained. The liquid in this large-storage gas-holder was commercial glycerin instead of the usual water, so that alteration of gas composition by solution in the liquid in the gas-holder should be reduced to a minimum. In the single experiment with pure oxygen this gas was supplied as a stream from a cylinder, bubbling through baryta before the respiration chamber.

All stated values of oxygen concentration in this paper refer to the make-up of oxygen mixture as supplied to the chamber. No reduction to the mean oxygen content of the chamber is attempted.

Our study of apples in nitrogen already reported in Paper II suggested to us the existence of some special relations between fermentation and respiration, which led us to take a connected complex view of the totality of respiratory metabolism and to introduce the conception of *oxidative anabolism*. The experiments with nitrogen revealed that our Bramley's Seedling apples were not permanently disturbed even by long periods of complete absence of oxygen, so they should prove very suitable for experiments with reduced oxygen supply.

In the present communication we shall investigate the effects of low supplies of oxygen, such as are provided by 3, 5, 7 and 9 % O$_2$, these being associated with one experiment in 100 % O$_2$. Our objective in this section of our work will be to establish as much as possible about the nature of the oxygen control of the rate of respiratory metabolism in the apple tissue. Certain difficulties and complexities of a fundamental nature are inseparable from this quest, and these will be sketched in a preliminary way in the later sections of this Introduction.

Then in Part I we shall proceed to collect facts by a direct observational survey of respiration under varying low supplies of oxygen. In the next paper in the series, Paper V, we shall manipulate these direct observations so as to provide evidence about the oxygen control mechanism, aiming in conclusion, in Paper VI,* at the metabolic interpretation of the general pattern of this particular control.

The first of the fundamental complexities indicated above is the fact that in 'low O$_2$'† both types of respiratory metabolism occur concurrently in the same tissue, so that there is produced a mixture of the CO$_2$ of fermentation symbolized by us as NR (nitrogen respiration) with the CO$_2$ of aerobic respiration, distinguished as OR (oxygen respiration). Observing the CO$_2$ production of the living apple we cannot make a direct allotment between the components but must adopt a collective term TR (total CO$_2$ production) so that TR $=$ NR $+$ OR.

---

* Papers V and VI were labelled by Blackman as Parts II and III to follow the present Part I. For publication here Parts II and III are called Papers V and VI respectively.

† The constant occurrence, in this paper, of references to oxygen and carbon dioxide, merely in a *general quantitative* sense, justifies the use of some abbreviated expressions; and we propose to employ the symbols 'O$_2$' and 'CO$_2$' where this quantitative sense is the only implication. Thus low 'O$_2$' will cover equally low amounts or supplies or concentrations of oxygen in connexion with respiratory metabolism.

A further fundamental difficulty that underlies a study of the pure control by oxygen is that we must recognize a second control of rate of CO_2 production in the grade of carbon activity, and that these two not only interact as limiting factors but also have other indirect relations. This matter must be sketched in some detail and the next section of this Introduction will be devoted to it, while the third section will deal with the problem of the best approach in order to bring out the exact oxygen control.

2. *The relativity of CO_2 production to 'oxygen condition'* *and to 'carbon grade'*

The continuous production of CO_2 at all states of plant respiration predicates sources of production of activated metabolic carbon as well as sources of production of activated oxygen. The rate of each is subject to a tissue control and their interaction must have some determinable pattern, which we shall aim at defining later in this paper.

The sources and controls of these two elements differ widely. We shall not seek here their individual biochemical transformation but shall endeavour to present our material in a generalized physiological terminology. The aspect of primary physiological importance is a kinetic one, namely, the *rates* at which active carbon and active oxygen are made available, both potentially and actually. As a brief general term for the actual rate of production of active carbon units at any moment we propose the phrase 'carbon grade'.

While the carbon grade of an apple at any respiring moment is the outcome of its own internal metabolism, the oxygen condition can be made to vary infinitely by the domination of the experimenter over the external supply of oxygen to the potentiality of the internal oxygen mechanisms.

A few points may be made about the oxygen factor before passing to a fuller sketch of the carbon factor.

Oxygen states of respiring tissues. This matter will be developed further in later papers. Ultimately, such an exposition must be made capable of dealing with different types of experimental respiration chambers as well as with the plant's whole range of respiring tissues. We may distinguish between the external and internal states.

(1) For the conditions external to the tissue we shall speak of the *external oxygen supply* to the tissues, a conception combining elements of rate of supply as well as of O_2 concentration in the experimental gas current, in addition to the diffusive magnitudes of the chamber's dimensions.

(2) For the internal state, we propose to speak of the *internal oxygen condition* of the tissues, combining the conceptions of the consumption rate achieved in oxidation and of the internal partial pressure of O_2 maintained in the cells, against the diffusion resistance of the tissue system.

With this brief introduction upon the oxygen factor we may pass to consider the characteristics of the *carbon metabolism* which the respiring apple tissue maintains for interaction with the internal O_2 condition.

The carbon grade of respiring tissues. Our sequence of experiments was carried on continuously through nine months, and during that time the rate of respiration of the stock of Bramley's Seedling apples, kept in cool storage at $2.5°$ C., underwent a systematic drift of intensity which was fully described in the earliest paper (Paper I).

This slow but considerable change of CO_2 production, carried out in spite of the external O_2 supply remaining constant, as air, all along, is taken as evidence of a deep-seated changing control of carbon metabolism, labelled by us the 'ontogenetic metabolic drift'.

In addition to this progression of metabolic stages, there is observed in the course of each individual experiment a general decline of respiratory intensity with time, as long as the apples are kept at $22°$ C., in the respiration chamber; and this decline manifests itself, whatever may be the concentration of oxygen supplied. As a first approach, and ignoring subtle equilibrational relations, this phenomenon is conveniently viewed as a *starvation* effect, involving declining production grades in carbon metabolism.

The result of experimenting with material subject to multiple carbon-metabolic change is that the rates of CO_2 production given by two applications of one definite O_2 supply may depart widely from one another, according to the ontogenetic stage and the starvation grade of the apple that is under experiment.

While the experimenter may maintain perfect control over the O_2 supply that is superimposed upon the tissues, we cannot yet in practice produce quick changes of the carbon grades that also control CO_2 production. The conditions of the inter-action of the carbon and oxygen factors must therefore be closely scanned.

The carbon and oxygen relativities as affecting CO_2 *production.* In all concentrations of oxygen, including zero, there takes place production of CO_2. We have no null point from which to start our ordinate scale of rate values. Further, as internal carbon variations may put CO_2 at a high or low level we cannot hope to get the same absolute magnitude of CO_2 rate always correlated with the same O_2 supply. We can only expect to be able to state *relative rates* of CO_2 production; which rates are to be defined as a ratio to the CO_2 in some one O_2 supply adopted as a unitary standard. Theoretically any arbitrary O_2 might be adopted as standard, but certain concentrations have a higher significance or greater convenience. Obviously the 21% O_2 of air is much the most *convenient* standard, as all apples had been examined in it. In this first survey of the experiments we have no hesitation in adopting the CO_2 produced in air as our unitary standard. If TR stands for the CO_2 in any low O_2, then the quantitative definition of respiratory rate for that experimental case must be derived from a ratio of the general form TR/AIR. Two independent values of CO_2 as TR and AIR are always needed for the formulation of the relative rate of R, in response to a change of oxygen supply, and these we may conveniently distinguish as experimental (of the numerator) and control (of the denominator or standard).

It is impossible to get both values concurrently on the same apple tissue, but if pairs of identical apples were a practical proposition the two values might be obtained simultaneously from them. As, however, it would require tedious preliminary experimentation to establish the metabolic identity of a series of pairs of apples, we have adopted a less exacting type of experimentation, carried out upon a single apple in a sequence of three stages (and so called 'a tripod experiment') which provides satisfactory evaluations of the relative rates.

With some plant tissues it may not be necessary to take all this thought about the two differing carbon-metabolic states of the tissues employed, but there is a striking character of apples which makes this refinement absolutely essential in their case. This character lies in the extreme slowness at which apples adjust their respiratory carbon metabolism after a change of O_2 supply. If values for TR and for AIR could be got, both within a few minutes of one another, then all would be simple; but the apple requires 30–50 hr. to adjust itself to a new state, during which time starvation, at an experimental temperature of 22° C., may have lowered considerably the carbon grade.

The procedure that we have adopted to meet this situation has been to carry out each experiment with a single apple kept for 50 hr. or more in its particular low O_2, between long periods in air. We have then to manipulate the collected data so as to distinguish between the three major factors at work as determinants of CO_2 production; planning to eliminate the carbon factors that are an expression of ontogenetic stage and starvation grade, while emphasizing the O_2 condition.

For the final section of our Introduction we may describe the ideal course of the experimental records all of which provide one or other of two clearly distinguished theoretical types of drift for the slow transition to the new adjusted state in low O_2. This will make clear the manipulation practised to eliminate such aspects of the carbon factor as are altered by factors other than the O_2 supply.

3. *Simple and compound transitions to adjusted and subadjusted metabolic states*

Our approaching exploration of the phenomena associated with oxygen conditions in apples will be primarily presented as a study of 'time-drifts' of CO_2 production. Such drifts may be either composed of observed TR rates or the *relative-to-air* ratios given by TR/ALR drifts, both plotted over an axis of time. These time-drifts are nearly always carried on for 50 hr. after a change of oxygen supply, since in less than that time one cannot be sure that the transition is over and that one has under observation a state which has become really adjusted to the new O_2 supply: nearly every time-drift displayed is, therefore, for most of its duration, only the tardy progression of a transition.

We recognize two main types of transitional drift which are easily distinguished from one another in any graphic record. These we label the *simple form* and the *compound form*. When the transition is over and the adjustment to the new oxygen supply is completed we can still make some minor graphic distinctions between the sequelae of the two transitional forms. After a simple transition we pass sharply into an *adjusted state* while after the compound transition we can only claim a *subadjusted* state. This latter has still some elements of decline in it which will be detailed later.

Postponing for the moment the description of the compound transition with its subadjusted finish we will first describe an ideal case of the simple transition to the adjusted state.

The simple transition and the adjusted state. The ideal record of such a 'tripod' experiment is presented in fig. 1, where, after a long time in air, the apple is supplied with a moderately low concentration of oxygen for some 75 hr. and then returned to air. The actual observed rate of CO_2 production follows the double line $A–B_1–B_2–D$. A represents the last value in air before the supply of low O_2, and is styled 'AIR *antecedent*'.

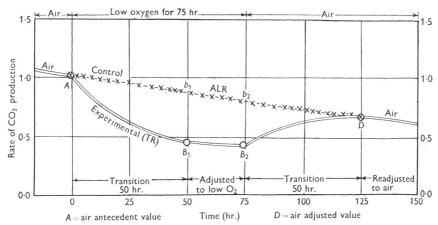

Fig. 1. Schematic presentation of the time-drift of CO_2 production through an idealized experiment on the effect of low O_2 supply for 75 hours between two periods in air. This drift is indicated by the double line $A–B_1–B_2–D$. Added to the figure by graphic interpolation is a drift line of crosses, which stands for the hypothetical 'air line of respiration' which gives a sequence of ALR values, $A–b_1–b_2–D$ along which it is assumed the experimental apple would have proceeded had it been kept in air all the time. This sequence of values would also be that given by the perfect 'control' apple in air to the 'experimental' apple in low O_2. This line of crosses supplies the denominator values for a sequence of ratios, TR experimental/ALR concurrent, that will express the *relative rate* of CO_2 production in low O_2. $A =$ last air value antecedent to low O_2; $D =$ first *adjusted* air value subsequent to low O_2.

This value, A, is followed, in low O_2, by the declining TR series curving down to B_1, which has the appearance of marking the end of the long transitional phase and here TR at last becomes adjusted to low O_2 conditions and continues thus to B_2. So it would continue indefinitely, but at B_2, air is readmitted and the TR values incline upward in transition back to air, not attaining readjustment until D, styled 'AIR *subsequent*'. After this there would follow declining adjusted AIR values till conditions were changed again.

Between points A and D, passing along the line through B_1 and B_2 we have a whole sequence of TR values of the experimental state ready to provide numerators for an appropriate sequence of control AIR values as denominators, if both were obtainable at comparable metabolic states, not distorted by individual peculiarities.

It is easy to interpolate graphically a curving line joining A to D by a track harmonious with the curvature antecedent to A and subsequent to D. This line would transverse the points $A-b_1-b_2-D$ in fig. 1, and provide a sequence of hypo-thetical 'air-line' values ALR, such as would have been observed could the experimental apple have been kept simultaneously in air as well as in low O_2. Air-line values will be symbolized by ALR so that their hypothetical derivation may be distinguished from actual observations in air, symbolized by AIR.

An ideal *control* apple, were it available, identical in ontogeny and starvation grade with the *experimental* apple, would, by definition, have given an observed drift in air that superposed on the antecedent AIR and subsequent AIR of the latter and followed the broken track ALR intermediately between A and D. Such an air-line has been constructed graphically for each experimental record and thus provides us with a sequence of ALR values concurrent with the sequence of observed TR values all the way from A to D. These concurrent ALR values serve as denominators to the observed TR values for the evaluation of the ratio (TR experimental/ALR control) which yields the *relative-to-air* rate of CO_2 production from moment to moment of time.

Investigation of the sequence of ratios thus obtained will provide an excellent implement for further probing into the low O_2 effects. Thus in fig. 1 the ratio would be 1·00 at A, and then decline in a series until, in due time (here first at B_1), the ratio attained constancy, marking just where the transition ended in the new adjusted phase, and our objective was attained. After this we could write:

$$\frac{\text{Low } O_2 \text{ experimental value}}{\text{AIR control value}} = \frac{\text{TR experimental}}{\text{ALR concurrent}} = \frac{B_1}{b_1} = \frac{B_2}{b_2}$$

= Ultimate adjusted ratio characterizing the low O_2 supply.

It is obvious that the starvation decline along the TR line may be so consider-able in 50 hr. that AIR *antecedent* is quite unsuited for a denominator to B_1 of the adjusted state, though it would have sufficed should adjustment to low O_2 have required only a period of, say, 1 hr.

The compound transition and the subadjusted state. The type of transition just displayed was described as simple because its graphic form suggests only a single equilibration system for the underlying mechanism, which system comes to a well-defined adjustment while under observation. There are, however, other transitional forms, suggesting dual metabolic mechanisms, and therefore to be described as compound.

In anticipation of the discovery of such forms to be encountered in Part I we may here give a condensed sketch of the characteristic features. The *compound form*, associated with change-over into nitrogen or into *very* low O_2, presents the special feature of starting off from zero hour with an immediate steep rise to reach an 'early peak' value of TR before the standard decline sets in. (An anticipatory glance at fig. 3, p. 97, will illustrate these features.)

In an attempt to extract a significant metabolic magnitude out of this complexity of initial form, we have applied the graphic technique of extrapolation of the curving drift, backwards from the 'peak' to the zero hour, thus arriving at an even higher initial value for TR.

We propose to regard such resulting *maximum initial* values of TR as providing fundamental metabolic magnitudes. All figures in the descriptive Part I of this paper which display an early peak have been given extrapolation. Details of values are given in the notes to the various tables.

We may add here a few sentences to indicate our line of justification for stressing this extrapolated maximum value.

The initial rise of TR here observed is held to follow very quickly upon the initial change of oxygen state of the tissues; working in the opposite direction, however, there is all along a metabolic factor inducing a decline in the rate of TR production. Were the change-over of the O_2 (and CO_2) states of the tissues accomplished instantaneously, instead of at the slow rate natural to the gaseous exchange of bulky tissues, then the TR peak, normally observed, would be anticipated by a series of still higher values declining from an instantaneous initial maximum of TR.

It is our adopted position that, as an index of the carbon metabolic state (carbon grade) of the tissues at the moment of change, the potential maximum is of greater quantitative value than the realized actuality, when the change is brought about in the presence of an independent factor of decline. As a first approximation, extrapolation from the peak may be used to provide a measure of the grade of the carbon metabolism of the tissues at which the change of oxygen condition is initiated in the apple, though the completion of this change may be delayed for some time by physical conditions.

We may now introduce a second feature that seems to be characteristic of the compound transition, and this is that the end of the transitional phase into a new adjusted state is not nearly so sharply defined as with the simple form of transition. In cases where we have had the patience to follow the adjustment beyond the standard duration of 50 hr. we observe that the TR/ALR ratio does not remain practically constant, period after period, but slowly declines giving the graphic feature that TR, if lying below the air-line is not convergent with ALR but diverges from it. For this state we propose to introduce the term *subadjusted state* without entering upon the question of the grade of departure from adjustment or what form the drift would have after the 100 hr., which has been the utmost limit of our inspection.

An example of a compound transition followed for 100 hr. is provided by apple XVI in 3 % O_2 (see fig. 3, p. 97) for contrast with XV in 5 % (see fig. 2, p. 92), which presented a transition that was practically a perfectly simple one.

This concludes our introduction which has, first, sketched the fundamental conceptions of respiratory metabolism involved in this line of research, and secondly, indicated the manner of collection and manipulation of data that will be employed to bring out the pattern of the oxygen control of the rate of this metabolism. In Part I we shall now examine the primary data, and in Paper V proceed to show how these can be manipulated to the desired end.

PART I. AN OBSERVATIONAL SURVEY OF THE CO₂ PRODUCTION OF
APPLES SUPPLIED WITH LOW CONCENTRATIONS OF OXYGEN

General introduction. In this observational part of the paper we propose to set
out and compare all the experiments on our apples V–XXV which involved the
use of low concentrations of oxygen as the oxygen supply. There are seventeen
cases, drawn from eleven apples, to be grouped into the four sections which make
up this Part I. The individual cases will be approached here in an exploratory
rather than an expository spirit, as it was in this way that they supplied the
foundations for the conceptions sketched in the Introduction, and the principles
formulated in Papers V and VI.

All these cases appear in the primary records that were long ago set out graphically (see Appendix I), and the excerpts from these records are presented here in a
form suitable for easy comparative survey. Tables 1, 2 and 3* were first constructed
by ruling up the large-scale graphic records and reading off the CO₂ values that
were to appear as TR and ALR in the first two lines of each table. For this purpose
a median track was drawn for the CO₂ values between the double lines delimiting
the tracks in the original graphic records (see Appendix I). In these records all
values of CO₂ were plotted as 3 hr. periods over one continuous time axis for each
apple. Now we need time scales which begin with zero hour at each change-over
from one oxygen supply to another.

The hours at the heads of the columns of the tables give the time-points selected
for ruling up the records so as to enable us to get out a definite form for the long
transitional drift and produce also values for the adjusted phase which succeeds it.

From the tables thus derived we proceeded to construct figs. 2 and 3 to illustrate
the text of this observational survey of the 'time-drifts' of CO₂ values, on change
of oxygen supply. Such time-drifts never show a level course, and they will be
said either to *decline* or to *incline*.†

These graphic figures have no numerical values on their ordinate axes, though
the CO₂ values of TR and ALR are recorded in the respective tables. The individual
figures are brought close together, not for study of their pitch but almost entirely
to bring out the *forms* of the transitional drifts. Reference to the tables will
reveal the considerable variations of pitch that are involved, arising out of
variations of ontogeny and starvation, referred to in the Introduction, p. 85.

With regard to the ontogenetic history of the particular apples that happened to
have been examined in low oxygen supply, it may be reported that none of them
belonged to class C, which differed superficially from A and B in its type of
behaviour in nitrogen. The apples with low oxygen supply belonged partly to
class A and partly to class B, but we cannot detect any systematic difference to
be associated with this labelling. This observational survey finds its completion
in the ratios which appear throughout as the third lines in the tables. Here we
have inserted for every time point the ratio between the first two lines. These

---

* These tables are at the end of this paper.

† The terms *rise* and *fall* of respiration, or *increase* and *decrease* of respiration, are to be
reserved for use in other connexions than the features of these time-drifts.

ratios of TR/concurrent ALR represent the *relative-to-air* rates of CO_2 production discussed on p. 88.

These particular ratios may be called 'air ratios', and their ultimate value when the tissues have become adjusted to the particular low O_2 state is a matter for discussion and determination in each of the experiments in this part of the survey. Subsequently, in Paper V these air ratios will form the starting point for new inquiries into the metabolic mechanisms.

4. *Respiration with an oxygen supply of* 5%

On six occasions apples were examined with an oxygen supply of 5%, and these occasions were well distributed as to their dates so that some of the individuals were in early metabolic stages while others were in the later senescent stages. The spontaneous ontogenetic metabolic drift of apples in air has been presented schematically in Paper I, p. 9, fig. 3, and this reveals that the first four apples to be examined in 5% O_2, numbers IX, XI, XIII and XV, were in the stages characterizing the ascending limbs of A or B of that schema while XVII was at the crest, and XX was well down the descending limb.

The respiration data of these six apples will be found in table 1, while the drifts of the TR and ALR values are set out in fig. 2. Each record begins with a short piece of broken line representing the antecedent AIR drift. The continuous line drifting on from the point of change of oxygen supply gives the TR drift, declining until a subsequent change of oxygen supply, to mark which there is drawn a terminating vertical line. The dotted line continuing from the AIR drift gives the sequence of interpolated ALR values appropriate to that particular apple.

Let us look first at the case of apple IX examined in December. In about 3 hr. after passing into an oxygen supply of 5%, the CO_2 production is clearly falling off, and the TR drift declines, steeply at first, and then gradually rounding off to a slighter declining straight line. It does not reach this feature until after about 30 hr. When we examine the air-line of IX we see that, after a level start, it also is declining, but somewhat more steeply than the last stage of the TR drift so that the two lines are now convergent. An obvious suggestion would be that TR and ALR are now maintaining a constant ratio to one another. Inspection of the ratios entered in the third line of the table shows that the rapid initial decline of the ratio TR/ALR *concurrent*, has given place to a constant value of 0·60 at about hour 30.

The observation of the graphic form of TR and of the ratios both combine to suggest that we have to face a transition lasting 30 hr., followed by an 'adjusted phase' (see Introduction, p. 88) characterized by a constant ratio.

In the case of apple IX the oxygen supply was changed back to air at hour 50. The figure shows us that a slow reversal of the respiration effect at once sets in and the time-drift of TR now inclines, at first steeply and then slightly, till after about 30 hr. it is judged to be completely readjusted to air, again having manifested a long-drawn-out transition. It appears that the past lowering of the respiration to 0·60 ALR is entirely reversible and without injurious after-effect.

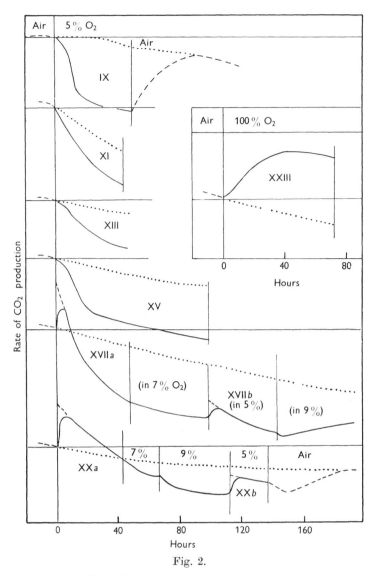

Fig. 2.

General note to figs. 2 and 3. All these records are graphic presentations of the TR and ALR values of the experiments in tables 1–3. The ratios in the third lines throughout the tables are reserved for graphic presentation in fig. 2 of Paper V. The contents of these figures are described in detail in the adjacent text of §§ 4–7; and they have been introduced by the schematic fig. 1 together with the comment and outlook on terminology in §§ 2 and 3.

The periods in low O_2 extend from zero hour to the next vertical line and the time drifts of TR values are represented by heavy continuous lines. Periods in air, represented by broken lines, precede the low O_2, and often follow it. The lines of dots represent the deduced ALR values expounded in § 2 as the values for apples kept continuously in air. Where the form of the TR drift exhibits an early peak, extrapolation back to an initial maximum value is represented by a broken line curve (see the general note to the tables).

Ordinate values are omitted in these figures and there is no indication of the positions of zero TR values. No visual judgment of TR/ALR ratios is therefore possible from these two figures. These ratios, which appear in tables 1–3 ready for discussion in Paper V are given graphic presentation in fig. 2 of that paper.

Special note to fig. 2. This figure contains eight time-drifts in 5 % O_2, and an inset figure for one apple in 100 % O_2. This latter is described in § 7.

For apples XVII and XX the two periods in 5 % O_2 are separated by periods in 7 and 9 % O_2. These intervening periods form the subject-matter of § 6.

This period of 30 hr. is the minimum duration observed for any transition, in other cases 40 or more hours have been recorded. We have therefore decided to adopt 50 hr. as a time at which the adjusted state may be safely assumed to have arrived and we shall use the TR/ALR *concurrent* ratio at 50 hr. as the standard 'characterizing ratio' of *relative-to-air* rate of CO$_2$ production for the new oxygen supply.

If the view is to be urged that the effect of a low oxygen supply can be simply summed up as depressing TR to a definite fraction of concurrent ALR, it is desirable to seek confirmation of this by carrying on the low oxygen supply for a much longer period than the first 50 hr. Apple XV was devoted to such an exploration and its values appear at the foot of table 1.

Apple XV was first examined for 144 hr. in air, so that the form of the AIR time-drift might be securely established, after which the oxygen supply was changed to 5%. The respiration declines in a drift that is hardly distinguishable in form from the record of IX, and in about 40 hr. it also proceeds along a gently declining linear slope. The 5% oxygen supply was maintained for 100 hr. in all, and there is no sign of change in the TR slope. As more than 10 days had been given up to this simple experiment, in its air and 5% O$_2$ parts, it was then brought to a close, though it would have been better had XV been returned to air once more for confirmation of the location of the air-line drawn in the figure, by extrapolation of the early part actually recorded in air.

The ratios for this case are to be found at the bottom of table 1. During the transition of about 40 hr. they decline systematically, but in the adjusted phase which follows they remain practically constant for some 50 hr. The tabulated ratio of 0·71 at hour 50 is the characterizing ratio to be adopted for XV and an oxygen supply of 5%, experimented on in February. There is noted a slight decline after hour 50, but the ALR denominator has not now enough authority for discussing such small movements of ratios.

This pair of experiments is held to establish the general form of the TR drift in the low oxygen supply, and also the division of the drift into transitional and adjusted phases, but we note for further exploration that the adjusted ratios differ considerably, being 0·60 for IX and 0·71 for XV.

We next consider apple XI, which was examined with 5·0% O$_2$ in January. Fig. 2 shows that the drift is, in its type, the same as those of IX and XV, though the declining TR follows a course which is more rectilinear than the other courses. This may be partly due to the steeper decline of ALR which occurs in this case. This apple was only kept for 45 hr. in 5% O$_2$ and then treated with another oxygen mixture. In table 1 we find the TR/ALR ratios following very much the same course as in XV; the transitional period is not over till hour 40, and we may adopt the ratio of 0·72 given at hour 45 as our equivalent for the adjusted ratio at the standard hour 50. We find that this January apple conforms to the ratio found in February for apple XV.

In apple XIII we have still another January apple lying chronologically between XI and XV. The drift of XIII when transferred from air to 5% shows in fig. 2 a declining form which is identical with that of XV, though the absolute decline of TR below the air-line in fig. 2 is not so great. Again we see the transition lasting about 40 hr. and a rectilinear slope then taking its place. Here the adjusted 50 hr. ratio is found to be 0·70.

We may use a comparison of apples XI and XIII to stress the value of our method of characterizing the respiration effects in low oxygen supply, not by observed TR values but by the relative rates given by the TR/ALR ratios. Looking at the values in table 1 we see that the whole set of TR and ALR values is pitched much lower in XIII than in XI. From the original records in Appendix I the explanation of this is quite clear. XI was of a higher initial value of respiration owing to its ontogenetic class and state as shown in fig. 3 of Paper I, while XIII has a lower initial value, as the same figure shows, and has, in addition, undergone 4 days longer starvation at 22° C. This has brought about a much lower 'carbon grade' for XIII than for XI, and it will be seen that TR values and ALR values are down to about two-thirds of the values for XI. Yet the TR/ALR ratios at hour 50 are 0·70 and 0·72 respectively, showing that the starvation grade differences have been well eliminated by taking this *relative-to-air* rate as our index of performance.

The four apples described so far all lie on the ascending limb of the schema of senescent drift (Paper I, fig. 3, p. 9). The earliest and lowest pitched apple IX indicates a ratio of about 0·6 for the depression in 5 % O_2, while the three later ones, with higher pitches of TR, all give ratios of about 0·7. The change of ratio from 0·6 to 0·7 suggested that an increase of ratio might be an empirical characteristic of senescent drift, and this is supported by the later examples that we have yet to consider.

We now pass on to apple XVII which was subjected to a variety of oxygen mixtures, but a glance at its full record in Appendix I, or its simpler presentation in fig. 2, does not suggest that the depression of its respiration is very different in 5, 7 or 9 %. A good deal of study had to be given to this record, which, unfortunately, was not returned to air at the end of the series of oxygen mixtures, and we have built up a detailed interpretation of it. The construction of this was facilitated by the fractional air-lines drawn right through the course of the graphic record in Appendix I. The parts that concern us here are the two exposures to 5 % O_2; the first, XVII*a*, was made at hour 47* from the beginning, and the second, XVII*b*, at hour 144*. The TR drift for XVII*a* in fig. 2 shows a marked difference of form from the types we have met so far, for initially the respiration actually shows a rise, and that a fairly steep one, in 5 % O_2. This soon turns into a typical decline cutting across the air-line steeply at hour 12, and gradually getting less steep but even at hour 48 when the supply of 5 % was changed for 7 % the decline has not become rectilinear, and does not show the expected form of a slope convergent with the air-line. On the ontogenetic schema of Paper I, apple XVII lies at the crest of the senescent drift, and clearly something has happened in association with this progress to differentiate it from the earlier examples; for the

---

* These 'hours' refer to the graphic figures in Appendix I where 0 hours marks the beginning of the experiment. The hours stated below refer to fig. 2, where the transition to 5 % oxygen is at 0 hours.

transition from air to adjustment in 5 % is no longer a *simple* transition but shows a marked early peak and therefore introduces to us the *compound form* of transition which we have briefly characterized in § 3 of the Introduction.

The ratio reached at hour 50 may be taken as 0·74. After the apple had been in 7 % O_2 for a further 51 hr., 5 % O_2 was again given (XVII b). The record shows a repetition of the special form with an early peak, after which TR settles down to a decline which is rather more like a line convergent with the air-line, at a constant ratio of TR/ALR. The ratio here falls lower than in XVII a, and reaches 0·71. We may conclude that the ratio value of 0·74 would not have been maintained constant in 5 % but would have kept on declining very slowly in the way held to be characteristic of the subadjusted state, defined in § 3 of the Introduction as occurring in association with the compound transition. This record serves as a forecast of the still greater divergence of transitional form shown by the next example.

Apple XX examined in May was a representative of the descending limb of the senescent drift schema, and so was in an advanced senescent stage. At hour 97*, 5 % O_2 was given to it for the first time (XX a), and later on, after exposure to 7 and 9 % O_2, 5 % O_2 was given again at hour 210* (XX b). The record of XX a in fig. 2 appears, at first sight, to have nothing in common with the records of the early apples IX–XV, for the respiration rises rapidly in 5 % and only declines very slowly from this high pitch. This record presents a striking example of the compound transition, and as the figure illustrates, and the data of table 1 demonstrate more precisely, the TR/ALR ratios put the relative rate of CO_2 production above 1·0 until hour 44 when the O_2 supply was increased. For the standard ratio for hour 50 to characterize this apple we may adopt 0·98.†

We register this high ratio for XX a but conclude that a lower ratio would have been reached in time, had this first exposure to 5 % O_2 been prolonged. Turning to the second exposure to 5 % O_2 we find that by the previous long treatment with 7 and 9 % O_2 the respiration is already down to 0·8 of the air-line when the 5 % O_2 is given. The record of XX b does therefore not show the marked features of XX a which started from air, though it rises a trifle initially. This second treatment with 5 % unfortunately only lasted 24 hr., so that we cannot say how long this ratio would have continued to decline in its subadjusted state.

In this case the apple was returned to air directly after the 5 % O_2, and the return transitional drift is interesting, for it no longer has the simple form shown by the return of the early IX, but presents the compound form of an initial decline before the rising curve sets in, which ultimately attains the air-line. In this apple we note that both transitions are compound; both air to 5 % and 5 % to air start by moving in the opposite direction to that of their subsequent course.

This ends our series of experiments with 5 % O_2 which brings to light a remarkable shift in the form of the records. It is fortunate that we find ourselves in possession of well-distributed examples, especially so in this, that had the intermediate form provided by XVII been lacking it would have been almost impossible to consider XX as the end of a continuous series which began with IX.

The change that comes over the time-drifts during the series is two-fold in

* See footnote p. 94. † See special notes to table 1.

nature. All along the chronological series, which, biologically, is an ontogenetic drift, the ultimate *relative-to-air* rates (adjusted 50 hr. ratios) are rising in spite of the uniformity of the 5 % external oxygen supply.

Further, as a second change, late on in the series, clearly visible in XVII, we have the arrival of the compound instead of the simple form of the transitional drift.

The conclusion from this two-fold change is that the ontogenetic shift of stages has a residual *direct* effect upon oxygen relations, that persists after the elimination of its 'carbon-grade effect' by the method of employing relative-to-air rates. This effect must be registered here for 5 % O_2, and will have to be taken up again in Paper V.

5. *Respiration with an oxygen supply of* 3 %

Since the pitch of CO_2 production of apples in nitrogen is not so very different from that in air, while there is a large depression of pitch in the intervening 5 % O_2, we next examine some cases in 3 % to see how this supply plays its part between 5 and 0 %. Our three apples examined in 3 % O_2 were XIII, XIV and XVI. The data for their time-drifts are set out in table 2 with graphic presentations in fig. 3. Their numerals show that the three were closely adjacent in chronology, but of the three, XIII and XIV were very low in their AIR respiration before the low oxygen supply, while XVI was very much higher.

It will be worth while to inquire into their ontogenetic and starvation characters as a test for the application of our technique of using TR/ALR ratios for eliminating as far as possible the effect of such differences in carbon metabolism.

XIII and XIV were both low down on the rising ontogenetic limb of class B (see fig. 3 in Paper I), with a mean initial AIR value of 14·4; XVI was high up on the rising limb of class A with an initial AIR value of 19·5. The difference of pitch between XVI and the mean of the two others is therefore very great, and this is not to be attributed to any variety of starvation experience, which was 46 hr. for the mean of the pair and 45 hr. for XVI. So much for their prehistory; and table 2 reveals that they come into our present survey with 18·00 as the antecedent AIR of XVI and 12·7 as the mean antecedent AIR of XIII and XIV; and with 13·8 as the 50 hr. TR value of XVI while 9·2 is the mean 50 hr. TR value of XIII and XIV.

Here then we can ask, as a test of our use of ratios for eliminating the distortions of ontogeny (and starvation), whether these divergent CO_2 values will give similar TR/ALR ratios. Inspection of the 50 hr. ratios in table 2 shows 0·84 for XVI and 0·83 as the mean for XIII and XIV. We may therefore accept this technique with considerable confidence and pass on to compare the details of the time-drifts of these three cases functioning with the low oxygen supply of 3 %.

When we examine fig. 3 we discover that XIII presents us with a well-marked example of the compound transition, TR rising at once to an early peak and then declining in a curve cutting across the air-line at about hour 18. After this the steepness of decline diminishes and it flattens out much as did this same apple when previously supplied with 5 %. We may take as characterizing ratio the value of 0·81 at hour 50.

The actual record was cut off at hour 45, when the apple was again supplied with air; after a few hours of decline the record inclines up to the air-line by a flattening transitional curve, but taking 40 hr. to reach adjustment.

Apple XIV provides us with an almost identical companion to XIII, and both were examined at about the same date. The form in fig. 3 is very similar though the decline is less steep, but then the same difference characterizes the decline of the air-line. (We recall that XIV has had 68 hr. starvation at 22° C., while XIII

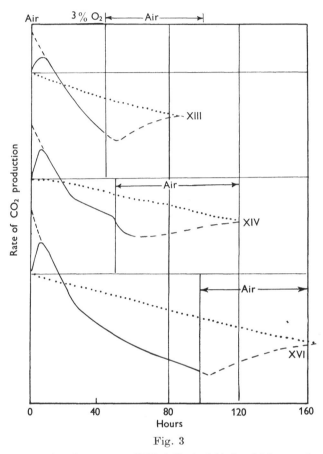

Fig. 3

Fig. 3 contains only the three cases of TR drifts in 3 % O_2 which constitute table 2. In all cases the drifts show a compound transitional form and are provided with initial extrapolation. Air is given after 3 % for all the three cases.

The general note set out under fig. 2 applies also to fig. 3.

has only been subjected to 25 hr.) TR cuts across the air-line in 18 hr., as did XIII. The drift of TR/ALR ratios in table 2 is much like XIII, and the characterizing ratio for XIV may be taken as 0·85, though there is not so much evidence of the ratios working down to adjusted constancy in 50 hr. as there was for the 5 % O_2 experiments.

After 48 hr. with 3 % this apple was returned to air and presents a more considerable initial drop in its transitional form than did XIII, on the way to readjustment in air.

Our last example in 3 % O_2 is apple XVI, and the distinguishing feature of this case was that XVI was held in this low oxygen supply for as long as 96 hr. in order that the later part of its time-drift might be made clear. At the start, it rises markedly to a peak above the air-line and then declines steeply to cut across it at 18 hr., as do the other two. After this it progresses in a particularly smooth form: the curvature passes over to a rectilinear course at about hour 52 and then follows a declining straight line which continues to diverge from the air-line. It becomes clear from this record that the metabolism is not attaining a fully adjusted state which involves a constant ratio of TR to ALR. Here the ultimate state can only be presented as one of subadjustment, as defined on p. 89. We adopt as characterizing ratio the value of 0·84 as recorded for hour 50. With further exposure to 3 % O_2 the ratio slowly declines and in hour 100 has fallen down to 0·74, but the ultimate fate in this low concentration of oxygen has not been explored. This decline does not seem to be due to any irreversible toxic action, for we see that when XVI was restored to air its TR curve inclined up to readjustment in air in absolutely the same way as XIII, which had only had a short exposure to 3 % O_2.

These three extremely concordant records in 3 % enable us to get a clear picture of the features that distinguish them from 5 % time-drifts.

Our next question must be whether 3 % O_2 provides an intermediate form between 5 % and nitrogen or a form still less like nitrogen than is 5 %. The answer to this is clear with one important proviso: that *the apples compared must be at about the same stage of ontogenetic drift*. The only stage we possess for 3 % is that of apples XIII, XIV and XVI, and it is clear that this stage can be taken as sufficiently adjacent to that of the 5 % cases XI, XIII and XV. In 3 %, as in nitrogen, we find the compound type of transition providing us with an *initial maximum*, which is absent in the 5 % apples that are under comparison.

When we compare, further, their adjusted states we find the same relation repeated; for the 50 hr. characterizing ratios are about 0·83 in 3 % as against 0·71 in 5 %, the former approaching nearer to nitrogen, which effect is characterized by the ratios above 1·0. Also, as in nitrogen, so the 3 % cases tend to reach only subadjusted states instead of the sharply adjusted states of 5 % cases.

Though the importance of 'stage of ontogenetic drift' was fully realized for the experiments on apples in air (Paper I), it might have largely dropped out of sight for low O_2 had we not collected such an extended series for 5 % O_2. Had we chanced to possess, for 3 % O_2, only case XIII, and for 5 % O_2 only case XVII, we must have reported that there was no essential difference between the two effects.

The simplest formulation of the situation is to say that, for any one given percentage of O_2, senescence shifts all the features of the time-drift progressively in the direction of the nitrogen time-drift; illustrating this with our evidence that the shift of features in 5 % is carried as far in XVII as to reach practical identity with the type of 3 % O_2.

6. *Respiration with oxygen supplies of 7 and 9 %*

We have now shown that CO_2 production is higher in 3 % O_2 than in 5 % O_2, marking a natural progression towards the still higher values in nitrogen. We want to know next whether CO_2 production with a somewhat greater oxygen supply than 5 % will show higher values to start the progress toward the air value or whether there may be yet found still lower values in this region before the inevitable increasing progress to the air values sets in. For this purpose a few experiments were made with 7 % O_2 and with 9 % O_2, as oxygen supply.

None of the experiments available in this particular region of low oxygen supply presents quite the simple lay-out that we had employed for 5 and 3 % O_2. No one of the four records shows air as the antecedent gas but all came from some other adjacent low O_2. Brought from air to middle low O_2 all the transitional drifts would necessarily have declined, but when brought from 5 to 7 or 9 % O_2 we cannot have the extensive changes we inspected in the previous sections. But we shall get enough guidance to enable us to determine the direction of drift for these small changes of O_2 supply, and also the simple or compound form of the transition.

Antecedent low O_2	Experimental low O_2	Resulting change of TR	
From 50 hr. in 5 %	To 7 % for 50 hr.	Slightly down	XVII
From 50 hr. in 5 %	To 9 % for 35 hr.	Dip and up	XVII
From 40 hr. in 5 %	To 7 % for 25 hr.	Definitely down	XX
From 25 hr. in 7 %	To 9 % for 50 hr.	Further down	XX

The experiments on 7 and 9 % were confined to apples XVII and XX, and they were rather mixed up with the pair of experiments on 5 % carried out on each of these apples, so it may be worth while to set out formally as a text-tabulation the time spent by the apples in their antecedent oxygen supplies as well as in the respective experimental O_2 supplies. This is the more desirable because two records fall short of the 50 hr. required for the adjustment and have had to be carried on by extrapolation. The lower part of fig. 2 contains the whole records of XVII and XX and shows the sequence of concentrations while the actual values of TR and ALR as well as the TR/ALR ratios are to be found in table 3.

The resulting changes of TR values are set out in the text-tabulation on this page, and we see that they vary both in direction of drift and in intensity; while some are simple in form others are compound. Yet in all cases the experimental O_2 supply is higher than the antecedent supply. The answer that is here provided to the question about 7 and 9 % formulated in the opening paragraph of this section must evidently be both multiple and conditional. For both XVII and XX the change from 5 to 7 % produces a drop in the rate of CO_2 production, indicating that the lowest values are located decidedly on the air-side of 5 % O_2. For XX a change from 7 to 9 % results in a further drop, so that the lowest value may be located as far from 5 as is 9 %. For XVII a change from 5 to 9 % gives a contrary result, for after an initial dip the TR rate rises to a higher level. This drift has then the 'reverse-compound' form (see p. 95) of the type that a change from 5 % to air would give. Clearly then the lowest value of TR is somewhere in the region of 7 % for XVII and located somewhere near 9 % for XX.

This variation for 9 % recalls the variations that were established between different apples for 5 % in accordance with the progress of ontogenetic senescence. When we inquire into the ontogenetic stage of our apples XVII and XX we recall that the former was placed at the apex of class A in the schema of Paper II, while XX came lowest down on the descending limb of this class and was therefore more highly senescent than any other apple. In accordance with this we find its lowest TR closer to air than in any other case. For the oxygen supplies of 7 and 9 % we lack any example from the early ontogenetic stages.

The ontogenetic shifts we have noted above in TR behaviour in the region of 5–9 % O_2 call for a more exact regional exploration, but this can be better carried out when TR/ALR ratios are invoked to provide a superior surveying instrument.

For the present observational inspection all we can do is to adopt the best possible 'characterizing ratios' for the standard 50 hr. points and set them out for use in the later analytical survey of § 3 of Paper V.

After 50 hrs. in...	5 % O_2	7 % O_2	9 % O_2
Apple XVII	0·74	0·71	0·81
Apple XX	0·98	0·85	0·77

These ratio values seem to be fairly acceptable on the evidence of their formal graphic relations as established by table 3, though their interrelationships as a system are not at present particularly obvious. We shall next refer to these values on p. 117, where the lessons of such ratios find general consideration.

7. *Respiration with an oxygen supply of* 100 %

After we had made many experiments with low O_2 we decided that one experiment should be added with a high oxygen supply, above that of air, so that we might at least know whether the principles we had found in the lower region were at work also in the higher region, and would be applicable to the whole of the phenomena. If there were no new disturbing principles at work then it would be desirable to get a value of the adjusted TR/ALR type in a high oxygen supply, and learn whether the rising values up to the AIR value of 1·0 continued their rising progress on beyond air. From this point of view we decided to examine apple XXIII in pure oxygen, using compressed commercial oxygen from a gas cylinder.

Apple XXIII was a fully ripe apple assigned to class B, and its full experimental record is in Appendix I. The graphic record of its change from air to pure O_2 appears as an inset figure in fig. 2, and the usual data are set out at the foot of table 2.

We may first look at fig. 2 for the form of the time-drift of CO_2 production following this big increase of oxygen supply. We find that TR at once starts to incline upwards, while the course of the ALR drift is markedly downward. The steep incline of TR gets slighter and flatter and, at about 45 hr., starts to decline again in a typical starvation slope. This transitional form is exactly what our work on low O_2 would lead us to expect. It indicates that the duration of the transition is about 45 hr., and, to check that, we have the test of calculating TR/ALR values and noting when the shifting ratios (here a rising transitional series) settle down to the steady value that should characterize the adjusted phase.

The ratio drift in table 2 places this at about hour 50, and we note that hours 60 and 70 maintain the same ratio. Everything here is according to expectation, though the 'characterizing' ratio has the high value of 1·40 for 100 % O_2. We learn therefore that ratios of this type must maintain a steady rising sequence from our lowest values right on through air to pure oxygen.

We have no experiments for this oxygen supply at any early stages of ontogeny, when possibly a lower ratio might have been recorded.

From this direct survey of the facts of respiration for a range of oxygen supplies we have collected a corpus of TR/*concurrent* ALR *ratios* which will form the basis of the survey and formalization to be carried out in Paper V in our search for the pattern of the oxygen control of respiration.

Table 1. *Values of* TR, ALR *and their ratios in* 5 % O$_2$

Apple number		Hours in 5 %											
		0	5	10	20	30	40	50	60	70	80	90	100
IX	TR	14·0	13·0	11·2	9·0	8·4	8·1	7·8					
	ALR	14·0	14·0	14·0	13·9	13·7	13·3	13·1					
	TR/ALR	1·00	0·92	0·80	0·65	0·61	0·61	0·60					
XI	TR	16·6	16·0	15·1	13·3	11·9	10·9	10·7					
	ALR	16·6	16·4	16·2	15·8	15·4	15·0	14·8					
	TR/ALR	1·00	0·97	0·93	0·84	0·77	0·73	0·72					
XIII	TR	10·0	9·6	9·2	7·8	7·0	6·5	6·3					
	ALR	10·0	9·9	9·8	9·5	9·45	9·2	8·95					
	TR/ALR	1·00	0·96	0·94	0·82	0·74	0·71	0·70					
XVIIa	TR	(22·6)	(20·9)	19·1	16·1	14·5	13·4	12·8					
	ALR	18·9	18·75	18·5	18·25	17·8	17·5	17·3					
	TR/ALR	1·20	1·11	1·03	0·88	0·81	0·77	0·74					
XVIIb after 7 % O$_2$	TR	(13·0)	(12·6)	12·15	11·4	10·8	10·4	10·3					
	ALR	15·8	15·6	15·45	15·2	14·9	14·6	14·5					
	TR/ALR	0·82	0·81	0·79	0·75	0·72	0·71	0·71					
XXa	TR	(14·8)	(14·2)	(13·6)	12·5	11·4	10·6	[10·2]					
	ALR	11·3	11·2	11·0	10·8	10·65	10·5	[10·4]					
	TR/ALR	1·31	1·27	1·23	1·16	1·07	1·01	[0·98]					
XXb after 9 % O$_2$	TR	(8·9)	8·8	8·6	8·4	8·3							
	ALR	9·8	9·7	9·65	9·5	9·5							
	TR/ALR	0·91	0·91	0·89	0·88	0·87							
XV	TR	15·0	14·5	13·7	11·25	10·3	9·9	9·6	9·4	9·2	9·0	8·7	8·5
	ALR	15·0	14·8	14·7	14·4	14·1	13·8	14·1	13·3	13·1	12·9	12·7	12·5
	TR/ALR	1·00	0·96	0·93	0·78	0·73	0·72	0·71	0·70	0·70	0·70	0·68	0·68

Table 2. *Values of TR, ALR and their ratios in 3 % and 100 % O₂*

Apple number		0	5	10	20	30	40	50	60	70	80	90	100
										Hours			
XIII 3%	TR	(16·1)	(15·0)	14·0	12·1	10·4	9·4	9·1					
	ALR	13·2	12·9	12·7	12·3	11·9	11·4	11·2					
	TR/ALR	1·22	1·17	1·10	0·98	0·88	0·82	0·81					
XIV 3%	TR	(16·4)	(14·9)	13·3	11·45	10·55	9·8	9·3					
	ALR	12·2	12·0	11·9	11·7	11·4	11·1	10·9					
	TR/ALR	1·34	1·24	1·12	0·98	0·92	0·88	0·85					
XVI 3%	TR	(22·5)	(21·0)	19·7	17·2	15·55	14·5	13·8	13·15	12·7	12·2	11·6	11·1
	ALR	18·0	17·8	17·65	17·35	17·1	16·75	16·4	16·0	15·7	15·4	15·15	14·9
	TR/ALR	1·25	1·18	1·12	0·99	0·91	0·87	0·84	0·82	0·81	0·79	0·77	0·745
XXIII 100%	TR	15·2	15·8	16·7	18·0	18·65	18·9	18·9	18·65	18·4			
	ALR	15·2	15·0	14·75	14·45	14·1	13·8	13·55	13·3	13·2			
	TR/ALR	1·00	1·05	1·13	1·24	1·32	1·37	1·39	1·40	1·40			

Table 3. *Values of* TR, ALR *and their ratios in* 7 % *and* 9 % O_2

Apple number		Hours						
		0	5	10	20	30	40	50
XVII in 7 % after 5 %	TR	12·8	12·6	12·25	12·0	11·7	11·5	11·45
	ALR	17·25	17·1	16·95	16·6	16·35	16·05	15·8
	TR/ALR	0·74	0·74	0·73	0·72	0·71	0·71	0·72
XX in 7 % after 5 %	TR	10·35	9·75	9·35	9·0	[8·8]	—	[8·35]
	ALR	10·35	10·25	10·15	10·05	[10·0]	—	[9·8]
	TR/ALR	1·00	0·95	0·92	0·90	[0·88]	—	[0·85]
XVII in 9 % after 5 %	TR	10·3	10·05	10·15	10·05	10·8	—	[10·9]
	ALR	14·5	14·4	14·25	14·0	13·8	—	[13·4]
	TR/ALR	0·72	0·70	0·72	0·74	0·78	—	[0·81]
XX in 9 % after 7 %	TR	8·8	8·3	8·0	7·7	7·6	7·5	7·45
	ALR	10·0	10·0	9·9	9·85	9·8	9·75	9·70
	TR/ALR	0·88	0·83	0·80	0·78	0·77	0·77	0·77

General note to tables 1, 2 *and* 3

Values of observed CO_2 production TR (except those in brackets) are derived as middle values between the 'contour lines' drawn on our large-scale graphic records of the individual experiments by ruling them up at 5, 10, 20 and more hours after the change of O_2 supply. These records are reproduced on a small scale in Appendix I, in which they can all be easily recognized.

The values of TR in round brackets are values deduced by our *initial extrapolation technique* to take the place of the observed TR values where the transition has a compound form. The intention and method of this technique are fully described in § 3, pp. 88 and 89, setting out the importance of getting *initial maximal values* of TR in such cases. The notes to the individual tables state the observed TR values which have been displaced by these higher values for each case of extrapolation.

The values of ALR have been read off from the single tracks forming the 'air-line' in the complete records of the experiments in Appendix I. The significance and value of 'air-lines' in relation to relative values of respiration is fully discussed in § 2.

The third line for each case sets out the TR/ALR ratios. These ratios are not shown in graphic figs. 2 and 3 of Paper IV, but are all collected into fig. 2 of Paper V for discussion in that paper.

Special notes to table 1. This table contains the values for the eight cases in which 5 % O_2 was employed. In four cases, XVII*a* and *b*, XX*a* and *b*, the initial extrapolation was called for, so the five actually observed values for hours 0, 2·5, 5, 7·5 and 10 are here set out: case XVII*a*: 18·9, 19·6, 20·4, 19·7, 19·1; case XVII*b*: 11·45, 11·9, 12·1, 12·2, 12·15; case XX*a*: 11·3, 13·0, 13·6, 13·7, 13·5; case XX*b*: 7·5, 8·4, 8·8, 8·7, 8·6. All but two cases were in air before the 5 % O_2; XVII*b* was in 7 % and XX*b* in 9 %.

The TR values given in the last column headed hour 50 were in several cases observed at the following slightly earlier hours, at which the experiment had been brought to an end: XI, hour 45; XVII*a*, hour 48; XVII*b*, hour 46. In XX*a* the 5 % ended at hour 44, but the decline there was still so steep that extrapolation had to be adopted to arrive at the values set out under hour 50; these are therefore enclosed in square brackets. Exp. XX*b* was cut short at hour 25, and extrapolation was not ventured on beyond hour 30.

Notes to table 2. See first the general note above. This table contains the three cases for 3 % and the one case of 100 % O_2. For all the 3 % O_2 cases the initial extrapolation technique was called for, giving the early values in round brackets. The observed values that these are based on were for hours 0, 2·5, 5, 7·5 and 10: XIII, 13·2, 14·3, 14·6, 14·4, 14·0; XIV, 12·2, 13·0, 14·0, 14·0, 13·5; XVI, 18·25, 18·7, 20·0, 20·2, 19·7.

The final value for XIII was actually at hour 45 and that for XIV at hour 48.

Notes on table 3. See first the general note above. None of these experiments was in air before its low O_2, and the actual history is given in column 1.

Only with XVII in 7 % was the duration of the low O_2 the full 50 hr. desirable, though XX in 9 % lasted till hour 46. XX in 7 % ended at hour 24, and the values at hours 30 and 50 are obtained by extrapolation. Case XVII in 9 % ended at hour 35 and the values for hour 50 are extrapolated, and so enclosed in square brackets.

V

THE ANALYSIS OF RESPIRATORY METABOLISM BY THE INTENSIVE STUDY OF 'AIR RATIOS'*

CONTENTS

INTRODUCTION

There is no aerobic or anaerobic condition of the typical higher plants in which the tissues do not continuously produce CO_2; there is therefore no state of zero production to serve as an origin for an ordinate axis of rate of CO_2 production. Accordingly, it is inevitable that some *standard* respiratory condition be established to which the rates in all other experimental conditions shall be stated as *relative* rates.

Our present standard of reference is that of AIR giving an external supply of 21 % O_2; the use of such *relative-to-air* rates has special advantages which we may now exploit, but also special disadvantages to be mentioned later.

These air-ratios are to be derived from the observational survey set out in Paper IV, and they have been recorded in the third line of tables 1, 2 and 3 of that paper as TR/concurrent ALR ratios. Physiologically they are conceived as relative-to-air rates under conditions of oxygen supply lower than 21 %, when all other metabolic conditions than those affected by O_2 supply are as near identical, in air and low O_2 as may be.

For a thoroughgoing survey of these air ratios we shall have to proceed methodically through the four sectional approaches indicated by the headings of the sections that make up this paper.

In § 1 we shall collect all our ratios and survey them as a family of transitions, displaying their drifts as a formal sequence determined by the supply of O_2 provided. In § 2 we shall discover that the complexity of the plant's organization is not countered by the application of this one principle of order, but that the sequence of ontogeny, leading towards senescence, introduces a new factor displacing the oxygen relations in a way that we shall have to define as accurately as we can. Then in §§ 3 and 4 we shall scrutinize the quantitative characteristics of special significant groups of air ratios that should help to bring us nearer to an elucidation of the pattern of the oxygen control of respiratory metabolism.

In Paper VI we shall move forward and set out our metabolic interpretation of the quantitative feature now to be established.

* This paper has not been published previously. No authors were named on the typescript.

1. The ideal Family of Drift Forms of the Air Ratios through Transitions to Low O_2

We settle down, first of all, to make an integrated study of all the forms of the time drifts of air ratios that have been recorded at transitions. Our primary observations of TR behaviour at transitions have shown us that the declining or inclining form will depend upon the direction and degree of change of O_2 supply starting the transition. In practically all our examples for this preliminary study the antecedent O_2 supply was air, the change was to a low O_2; and the period of observation was continued for 50 hr., in order to arrive at the air ratio of the ultimate adjusted state.

An extension of our survey will follow when transitions between two very low O_2 supplies come to be expounded (see p. 114).

The arresting observation has been made that the graphic forms of these drifts of air ratios vary according to the concentration of low O_2 which succeeds the air. In our collection of such forms we have already distinguished two classes, the simple class and the compound class of drift-form.

We shall now try to advance beyond this class and show that there is a definite family progression in the forms, associated with the grade of lowness of the experimental O_2 supply employed; the limiting cases being, on the one hand, continued persistence in air and, on the other hand, sudden change into pure nitrogen.

Our whole observed collection of forms is set out in fig. 2, but before dealing with its details we shall present fig. 1 which contains a schematic full sequence of these forms, arising on change from air to varieties of low O_2.

The forms could not be arranged on the basis of any single metabolic feature progressing step by step in one direction. Rather, there seemed to be a marked discontinuity of internal oxygen condition, located somewhere in the middle range of low O_2 supplies. We propound the thesis that there is a *critical concentration* of O_2 at which the break in continuity is determined by the establishment of a particular internal oxygen condition in the apple tissues.

It is therefore desirable to use the labels supercritical and subcritical to distinguish the oxygen supply and conditions which come above and below this critical supply.

For fig. 1 we have drawn schematically a selection of six forms of drift of air ratios labelled (a)–(f), each lasting 50 hr., which may serve to illustrate the details of the progressive sequence. For clearness, these are at first drawn in two separate groups. Form (a) stands for the limiting case of persistence in air, and here by definition the air ratio must present a straight line of the value of 1.0 throughout. At the other end, we reach at present to the form (f) for the transition into nitrogen. The intermediate forms (b)–(e) stand for arbitrary lessening supplies of O_2, (c) being attached to the special critical form.

Let us proceed through the complete sequence of changes of form, following on from form (a). Form (b) we think of as brought at zero hour into a definitely lower concentration than that of air, and there follows the characteristic slow transitional decline of air ratios, curving down to, say, 0.80 as an ultimate adjusted characterizing ratio for the new low O_2. Making a new start from air with a still deeper drop in O_2 supply we get, as form (c), a steeper transition form, declining to a still lower ultimate ratio, here 0·60. Such a falling progression of forms, however, will not be produced indefinitely, but will be arrested at some particular low concentration, which thereby becomes qualified to be termed the *critical* O_2 supply. With a somewhat lower O_2 there will arise such a form as (d), showing that a new sequence of changes is initiated. Henceforward, in strong contrast with former happenings, we are presented with a rising sequence of drifts, similar to one another but lying at higher and higher pitches, forms (d)–(e), as the percentage of O_2 is reduced, and ending with form (f) in nitrogen, where the air ratios lie above the air-line, all along the drift.

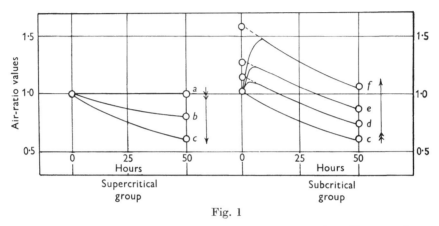

Fig. 1

Note to fig. 1. A selection of six arbitrary drift-forms lettered (a)–(f) to illustrate the complete sequence of transitional forms when these are built up of air ratios (TR/ALR concurrent). The 'critical form' (c) divides these into two groups (which, for clearness, are drawn isolated from one another), the supercritical (a) and the 'fermentation form' (f) standing for the transition into nitrogen. The numerical values adopted for the initial ratios (hour 0) and the ultimate ratios (hour 50) for all the six curves are set out here, below. The initial dotted regions of the subcritical curves indicate the parts that are derived by extrapolation, see pp. 88 and 89. The two arrows indicate the directions of progression of the ultimate air ratios with lower and lower O_2.

Forms lettered in fig. 1	a	b	c	d	e	f
Initial air ratios	1·00	1·00	1·00	1·12	1·26	1·54
Ultimate air ratios	1·00	0·80	0·60	0·80	0·90	1·1

The sequence (a)–(c) shown on the left hand of fig. 1 constitutes the supercritical group, where the transitions have the simple form while the four to the right constitute the subcritical group where occur the transitions described as compound, their observed 'early peaks' having been extrapolated by broken lines into initial maximal ratios (see pp. 88 and 89). It will be seen that form (c), that of the critical concentration, is really not a separate entity; but is a limiting case which in this drafting occurs in both sets and may be said to unite and divide the two groups of the progression. Were all the forms combined into one figure, form (c) would, of course, only appear once.

The three formal dimensions of air ratio drifts. We have now to pass from considering the general graphic appearance or forms of the transitional drifts to deciding what are the significant dimensions that show characterizing quantitative alteration from form to form through the complete sequence. As drawn in fig. 1 each drift declines smoothly from its initial air ratio to its ultimate air ratio, 50 hr. away, so that a pair of formal dimensions, one initial and one ultimate, stated in terms of the ordinate axis (air-ratio values) would serve for vertical dimensions, while the third dimension would be the duration between the two ends in terms of the horizontal time axis. We may now consider the variation of these three dimensions, one by one, through the sequence of forms. The magnitude of the two vertical dimensions would be the ordinate values of the six forms at the start and finish of the drifts.

The ultimate air ratio is the dimension of independent importance, because it expresses the new physiological state, adjusted to its particular low O_2 and, after the 50 hr. for full adjustment, this dimension has become quite *independent of the antecedent O_2 supply*. This state is also to be maintained indefinitely as a steady physiological state after the transition is over. Considering this dimension through our complete sequence of forms in fig. 1 we see that with form (a) this ultimate air ratio must have the value of 1·0, and that with lower and lower O_2 the air ratio would fall lower and lower, passing through form (b) assigned arbitrarily an ultimate ratio of 0·80, until the concentration of O_2 was reached giving the critical form (c) allotted an ultimate ratio of 0·60.

Passing on into the subcritical region with progressively lower O_2, the direction of change would reverse and the ultimate air ratio increase rapidly, being represented as passing through values of 0·8 for (d), 0·9 for (e), reaching up to 1·1 in nitrogen, definitely above the value in air.

From these ultimate dimensions we may now pass to the initial dimensions represented by the initial air ratios. The initial dimension consisting of our extra-polated, initial, maximum air ratio is in every way a contrast to the ultimate dimension. Let us first note its changes as we pass along the complete sequences of forms. With form (a) in air the value would of course be 1·00. Considering the sequence of lower and lower values in the supercritical group we see that this initial air ratio keeps, in the next two forms (b) and (c) of the progression, this same value of 1·0. It is only when we progress into the subcritical group that this initial dimension shows change. In the sequence of forms (c), (d), (e), (f) the initial air ratio shows higher and higher values allotted the intermediate values 1·12 and 1·26, until in nitrogen we have put it as high as 1·54.

On comparing the two sequences of changes shown by the ultimate and the initial air ratios, we can only conclude that these changes are determined by the operation of more than one factor. While the ultimate dimension of the air ratio is determined only by the new 'experimental' oxygen supply and should be the same, when the 50 hr. of transition have run their course, whatever had been the antecedent oxygen supply, yet the initial dimension is in part determined by both the antecedent and the experimental supply. In the present section we restrict ourselves to air as the antecedent supply, but in the next section (p. 115) we shall consider some cases with antecedent supply of low O_2.

Having characterized the changes of these two ordinate dimensions through the supercritical and subcritical regions we can now give a clear definition of that intermediate form which we have called the *critical form* of the drift of air ratios. The critical form must have an initial air ratio of 1·00, while its ultimate air ratio must be the smallest obtainable with that particular respiring tissue. As we have no method of forecasting the particular O_2 supply associated with such a form of transitional drift, our position must be that the supply of O_2 to be called the *critical supply or concentration*, in our particular oxygen-supplying apparatus and these special tissues, is that concentration of O_2 that does in fact give a transitional drift with the characteristics of the critical form, as just defined.

Employment of a slightly lower concentration supply than the critical in an experimental test will be found to raise both initial and ultimate dimensions, while a slightly higher concentration will raise the ultimate only and leave the initial as 1·00. If we know only that the ultimate is raised, and lack knowledge about the initial, then we cannot decide whether we have shifted our new form to the sub-critical or supercritical side of the original critical form. Further knowledge may, however, be gained by the technique described on p. 115.

The rate of 'adjustment' as the third dimension of transitions. The third dimension, that of the duration along the time axis, is concerned, physiologically, with the rate at which the initial dimension passes over into the ultimate dimension, in the process of adjustment. This dimension is of quite a different type from the other two, and we shall not discuss any range of variation in its rate for this first survey. Later, however, we shall have to show that the two components of the total compound transition show significant differences of adjustment rate.

The slowness of adjustment is an outstanding feature of the whole of this research upon oxygen concentrations. The time required may be as much as 30–50 hr., to the great delaying of progress in investigation. This slowness is so great that it cannot be attributed to physical delays such as diffusion and solubility resistance; its control passes out of the category of 'gas-exchange' phenomena. We can only conclude that there is a carbon metabolic transition which has also to be accomplished. Change of O_2 percentage initiates a change of rate of some metabolic feature, but this does not attain its adjusted ultimate balance of pro-duction and consumption of carbon metabolites until long after the balance was first disturbed.

Whether oxygen concentration in itself has a direct effect upon the *rate* of adjustment we cannot say for certain, but there is no obvious sign of such an effect. We may mention here the special phenomenon which we have called subadjust-ment (see p. 89), expressing itself in slowly falling ultimate TR/ALR values even after 50 hr. in very low subcritical concentration of oxygen so that there is no longer a sharp adjustment to be observed. Technically this might be viewed as a decreased rate of adjustment, but we cannot now delay to discuss whether it is largely a *toxic depression* of TR rate.

2. The Observed Air-ratio Drifts: the Shift in their Oxygen Relations caused by Ontogeny

In fig. 2 we have collected together all the transitional drifts of the ratio TR/ALR *concurrent* that are provided by tables 1–3 of Paper IV. They include air-ratio values ranging from about 1·5 to 0·6; all the drifts are plotted to a common ordinate axis of ratio values and are aligned to a common zero hour.

We have to survey their relations to one another and see that the correlation of sequence of dimensions of the drifts to O_2 indicated in fig. 1 (p. 108) are here justified.

The upper part of the figure contains only cases in which the apples were antecedently in air, while the lower part sets out the four exceptional cases which were changed from one low O_2 to another (for details see p. 99).

In the upper part all except one of the drifts decline, being changes into low O_2, while XXIII inclines steeply, being a change into pure oxygen. The thirteen declining drifts distribute themselves between the two classes of supercritical and subcritical, contrasted in fig. 1 (p. 108). To complete the survey of low O_2 drifts, four selected cases of transitions from air into zero O_2 have been brought into the figure from our investigations on nitrogen effects. These drifts, XXI*a*, XXI*b*, VII and XI, have been derived from Paper II (p. 42).

The supercritical class provides the simple form of transition, of which we have not many examples, as less attention was paid to this form than to the provoking compound form. In this class we look for drifts beginning with an initial air ratio of 1·00 and declining continuously to lower values for the ultimate air ratio. Apple IX in 5 % fulfils this and declines from its initial 1·00 to such a low ultimate air ratio, 0·6, that it will later be judged to represent a drift of the critical form, (*c*) in fig. 1, which should be characterized by having the minimal ultimate ratio. Apple XXIII, with inclining drift, was changed to 100 % O_2 from air, reversing the usual relation of concentrations, so that its inclining transition provides a fine example of the supercritical type presented the reverse way up, as the change was to higher O_2.

Passing to the region of subcritical O_2 we find that we have a wealth of examples. We shall expect to find drifts here corresponding to the forms *c–d–e–f*, as set out in the right-hand group of fig. 1. Accordingly, we see at the top of fig. 2 a sequence of examples with initial air-ratios ranging up from 1·0 to as high as 1·5.

It may be mentioned once more that these very high *initial* values are not observed values but are due to extrapolation of the observed 'early peak'; the exact changes thus produced are detailed in the notes to tables 1–3 of Paper IV.

It is clear from the general appearance of this figure that the individual drifts are arranged here like those in fig. 1 and do not intersect one another. As in the schema, the drifts representing transitions into nitrogen (form *f*) come into the top ranges of the figure, which claims all the forms of high dimensions, while below them, both in initial and ultimate values, come drifts in very low subcritical O_2 (say form *e*) represented by the group of 3 % curves, XIII, XIV and XVI. Next comes a group represented by XI, XIII and XV all in 5 %, which we would

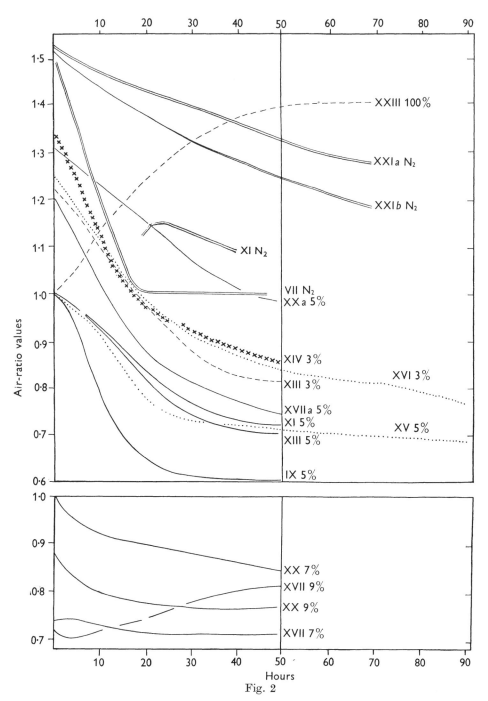

Fig. 2

Notes to fig. 2. All the observed transitional drifts following change into low O_2 at 0 hour are here presented as 'air ratios'—TR/ALR concurrent. The drifts are observed up to 50 or even 100 hr. so that the new adjusted state has been reached. All data are to be found in tables 1, 2 or 3 of Paper IV, except the nitrogen data which came from Paper II, table 1, p. 42. Each curve is labelled with the serial number of the apple and the percentage oxygen supply. The drifts in the upper part of the figure were all from air, antecedently, to low O_2 and to nitrogen and are discussed on pp. 111–14.

The four drifts in the lower part of the figure came antecedently, not from air but from some low O_2 concentration which may here be noted as being 5 % for the cases XVII 7 %, XVII 9 %, XX 7 %; while XX 9 % came from 7 %. Extrapolation had to be adopted to extend two of these drifts to the required 50 hr. Square brackets here and in table 3 indicate these derived extensions. These four drifts are discussed in the subsection on senescent shift of the critical concentration of O_2 § 2, p. 116.

associate with form (d)* as low subcritical O_2. The sequence would end with case IX in 5 % O_2 which gives the critical form (c), already described as the limiting case between the supercritical and subcritical groups.

We must now examine the progression of forms in fig. 2 more closely to see whether it presents any exceptions to this generalization that the sequence of forms is determined by oxygen supply.

Ideally each lettered drift-form would be associated with one particular concentration of oxygen throughout the experimentation; but this simple state is only partly realized. We have therefore next to explore the nature of the disturbing factor.

If we examine the cases in 3 % O_2 (XIII, XIV and XVI) we shall find that all three give practically identical drifts, of the form we label (e), so that the correlation here appears very close.

But if we next examine the six examples in 5 % we shall get quite a different result. In 5 % O_2 we find that IX, the lowest of all the drifts in fig. 2, has been labelled (c), while the next three cases, XI, XIII and XV, lie apart from it as a higher batch, showing form (d). The next case, XVIIa, comes higher still by itself, as form (e), while the highest, XXa, is even above the group of 3 % drifts, and we should now label this as intermediate (e) and (f). For 5 % then the examples range over forms from typical (c) to beyond (e), which is in complete contrast with the small range of the group of 3 % forms.

As the rising progression of 5 % drifts runs strictly in accordance with the chronological numbering, which is an index of the passing of time, bringing on a succession of ontogenetic stages, we can only conclude that it is ontogenetic progress which determines the progressive alteration of the observed forms of the drifts in spite of identical low O_2. A glance at the 3 % group brings out that all the chronological numbers are close together, while Appendix I gives dates which show that these three cases fall within a range of 3 weeks in contrast with the 4·5 months over which the 5 % set ranged.

Our conclusion must be that oxygen effects (even when cleared of carbon-grade effects) are not throughout independent of ontogeny, and that the later senescent stages of ontogeny can cause serious dislocation of the associations between transitional forms and low oxygen supply. The general direction taken by this displacement is clearly such that it may be said that for a given external low O_2 supply senescence brings about a higher value of the air ratio than was found at

---

* What has just been said about the representatives of form (d) now requires a little expansion, as it is in one way out of harmony with the drifts as actually drawn in fig. 2. Form (d) demands initial air ratios above the value of 1·0, while, in the figure, these three cases are all drawn as starting at 1·0. They have ultimate air ratios decidedly above the form (c) of IX, as would be demanded, and they can only be established as having the (d) form by giving to them higher initial air ratios also. On careful inspection of the start of this group of (d) forms in fig. 2, it is clear that they get on to their downward course much more slowly than does the drift of IX. We therefore suggest that this differential feature really represents a very small 'early peak' for the group of three, and that closer measures would have given distinct evidence of the peak reaching perhaps above 1·0, and therefore yielding on extrapolation an initial maximum, possibly as high as 1·1. If this line of correction of the figure is legitimate then we can discover, in this triad of drifts, a typical group of form (d).

the earlier stages of ontogeny. Such an increase in the value of the ratio might obviously be interpreted as due either to a decrease in the denominator representing CO_2 production in air or to an increase in the numerator TR, which is the sum of the fermentation NR + the aerobic OR in the particular low O_2. All these suggested effects might combine to contribute to the senescent rise of the ratio, but as a first expectation we may consider only the proposition that senescence has a general depressant effect upon the efficiency of *any* given external supply of oxygen.

Such a depression might be due to biophysical or biochemical changes in the respiring tissues, either to an increase of diffusion resistance to oxygen entry so that 5 % became, with senescence, only as efficient as 3 % O_2 had been formerly; or the senescence of the protoplasm might produce a deactivation of catalysts in control of respiratory metabolism which differentiated against oxygen activity relative to fermentation activity.

This analysis of primary causes may be postponed until all our observational survey is completed.*

For tracking the course of senescence through this complex of transitional forms and oxygen supplies there is one point which, both in interest and in accessibility, stands out as a suitable index of change. This point is provided by the drift of that form which we have called the critical form (*c*). This form locates the break between the two classes of transition, the simple and the compound, and therefore is an index of profound change of metabolic state. It is also more easy to characterize by experimental data because here there occurs, by definition, the *minimal* value of the ultimate air ratio for any given ontogenetic state.

The most illuminating query that we can set ourselves in this matter concerns the definition, for any senescent stage, of the particular grade of oxygen supply which gives rise to the transition with the critical dimensions. What we must seek to know is how this 'critical concentration' of oxygen shifts with senescence. As will be seen later (fig. 4 and table 2), for our apples, the critical concentration rises from 5 % through 7 to 9 % O_2. There is no evidence offered, however, that this is more than a middle selection from the total range of oxygen supply through which ontogeny might drive the critical concentration for this particular tissue.

The shifting critical supply of oxygen and the technique for locating it along the oxygen axis. The data set out in Paper IV make it clear that it is a slow process to obtain the well-established ultimate air ratio for a single preselected oxygen supply. Still more difficult would it be to obtain a record that ended in a preselected ulti-

---

* It will be appropriate here to justify our detailed attention to this senescent displacement, by evidence that its magnitude is considerable. We may pick out some values to illustrate this, using ultimate air ratios as an index. Let us take as our starting point for a comparison of the effect of ontogeny with the effect of lowered O_2 supply, as both being causes that raise the ultimate air ratio. For apple XIII, which gave the ratio of 0·7 in 5 % O_2 at the end of January, we found that the ratio could be raised to 0·8 by transference of this apple, at the same date, to 3 % O_2. However, apple XX*a*, in 5 % O_2 in the month of April, gave an ultimate air-ratio as high as 0·98, as against the 0·7 value in January. So we conclude that 3 months of increasing senescence in cool storage has been much more effective in raising the ratio than an immediate change from 5 to 3 % O_2.

mate ratio, so that one is led to think of how knowledge can be attained with the easiest technique of making approximations.

The easiest procedure for making an approximation to the critical concentration is to associate with determinations of ultimate air ratios knowledge of the direction of slope by which the transitional air ratio drift approaches this ultimate value. This enables us to decide between the ambiguous alternative and get some measure of nearness to the critical concentration.

This technique can valuably be applied to experiments which are not of the standard type, i.e. those defined in this survey by having air for the antecedent gas to low O_2, but are of a type in which the antecedent gas is some other low O_2 not too remote from the suspected critical region. Four such experiments based on change from one low O_2 to another are set out as drifts of air ratios in the lower part of fig. 2.

We may first consider the principles involved in this technique and then pass to the manipulation of the actual experimental records.

We set out in tabular form, in the text below, the canons that enable us to distinguish the four possible directions of transitional drifts in their relation to the locus of the critical concentration. Each line of the table begins with a definition of the experimental change of conditions that has been employed in interrogating the low oxygen region, stating whether the O_2 supply has been raised or lowered to bring on a transitional drift to the new adjusted ultimate air ratio. Then follows a statement as to whether the record of the transitional drift shows that the values have moved by a decline or an incline towards the ultimate air ratio. From the four combinations of these two factors we can deduce the resultant physiological direction of the alteration that is taking place. Two aspects of this deduced resultant are set out in the tabular form, one being whether the pair of states are on the supercritical or subcritical side of the critical concentration, and the other whether the new state is closer to the critical state than was the antecedent state, or further from it.

Oxygen change	End slope of drift form	Deduced direction of transition in relation to locus of (c)
(1) Lower O_2 supply	Decline to ultimate ratio	Towards (c) from supercritical side
(2) Lower O_2 supply	Incline to ultimate ratio	Away from (c) on subcritical side
(3) Higher O_2 supply	Incline to ultimate ratio	Away from (c) on supercritical side
(4) Higher O_2 supply	Decline to ultimate ratio	Towards (c) from subcritical side

Working by cumulative evidence in this way it is possible to piece together information which decides approximately the location of the critical concentration, provided of course that the ontogenetic stage is steady.

There is one type of 'slope of drift form' that is not provided for in the four tabulated cases, and that is where the observed drift keeps practically at a level value through transition, though O_2 has been raised or lowered. Such a rare exceptional state would be realized when the ultimate ratios of the antecedent and the experimental O_2 formed a pair, one on each side of the critical locus, but both of equal adjusted air ratios. The deduced drift then would be movement 'across (*c*)' from subcritical to supercritical or vice versa.

Let us now glance at the evidence we should propose to collect about location of critical concentrations of O_2 from the four drifts figured in the lower part of fig. 2 (p. 112). With regard to apple XVII, changed from 7 to 5 %, there is slight general decline; this corresponds to the tabled line (4), so we conclude that we are on the subcritical side and that 7 % is nearer the critical value than 5 %. The other test with XVII was to change from 5 to 9 %, and this showed a marked incline; this is the case of line (3) and indicates that 9 % is up on the supercritical side well beyond (*c*). Here then the critical value is about 7 %, and in the last change the apple was pushed over from the subcritical to the supercritical side.

Apple XX represents a more senescent stage than XVII. One experimental transition is again from 5 to 7 % O_2 and shows a steep decline. This indicates line (4), as with apple XVII, and shows 7 % is much nearer the critical value than 5 %. Finally, for XX the other transition was from 7 to 9 % with again a decline. This pushes the critical value right on past 7 % to somewhere about 9 %. On these pieces of evidence we decide that while the medium senescent stage XVII may have its critical locus at 7 %, that of XX has been shifted as far as 9 % and possibly further by the advancing senescence.

The senescent shift of the air ratios and of the drift form with uniform O_2 supply is of great interest from the point of view of metabolism and the biological status of the tissues, but it opposes a great resistance to the progress of the investigator who aspires to collect comparable curves for a variety of oxygen concentrations. In theory, for such a collection, all data would have to be carried through simultaneously on identical ontogenetic stages. This could hardly be done without the collaboration of a number of experimenters each with separate elaborate apparatus.

3. The Relative-to-air Rates of CO_2 Production in O_2 Supplies Lower than Air; the Ultimate Adjusted Air Ratios

An interesting approach to this matter is made by asking ourselves what the facts of this research upon apples would have looked like, had it happened that the transitional metabolism on change of O_2 supply had adjusted itself to the new state at a very quick rate instead of an inordinately slow rate. We are to suppose that all mechanisms of transition would still have moved to the same appointed ends but at such a rate that in a very short time the fully adjusted state for the new low O_2 would already have arrived, even already by the time that the first new trustworthy estimation of CO_2 production could be made.

We should, in such a case, have no knowledge of transitional forms or initial air ratios, nor should we need to make corrections for starvation grades or be forced to use different ontogenetic stages and afterwards make allowances for them.

The one type of value that would persist as having permanent significance would be a set of *adjusted ultimate air ratios*, but now bearing the alternative functional title of *rates of* CO_2 *production in low* O_2 *relative to that in air as unity*.

These ratios are a direct attribute of low O_2, and are entirely independent of the particular antecedent gas history, provided enough time has elapsed for drift right through the transitional states. It is upon the pattern of our set of these ratios, abstracted from their transitional antecedents, that we have to concentrate in this section.

We should at least expect these values to be largely under the control of oxygen supply and, also, to show ontogenetic distortion if the study ranged wide enough. Our obvious approach to the subject will be by collecting all such values into one table.

Table 1 presents all our '50 hour air ratios' drawn from the values in tables 1, 2 and 3 of Paper IV, which deal with our present experiments in low O_2, reinforced by a group of air ratios for zero concentration of oxygen. In Paper II (p. 42) the NR/ALR concurrent ratios were tabulated as transitional drifts for all of our apples V–XXV in which the behaviour in nitrogen had been investigated.

Table 1. *Relative rates of CO_2 production in low O_2 in the adjusted states, as represented by the ultimate air ratios* (TR/ALR concurrent at hour 50)

Date	Curve	Air initial	Apples	0 %	3 %	5 %	7 %	9 %	21 %
1 Dec.	1	12·8	VII	1·00	—	—	—	—	1·0
15 Dec.	1	14·0	IX	—	—	0·60	—	—	1·0
24 Jan.	2	14·2	XIII	—	0·81	0·70	—	—	1·0
24 Jan.	2	14·6	XIV	—	0·85	—	—	—	1·0
11 Jan.	2	16·8	XI'	1·06	—	0·72	—	—	1·0
10 Feb.	3	19·5	XVI	—	0·84	—	—	—	1·0
10 Feb.	3	20·5	XV	[1·11]	—	0·71	—	—	1·0
2 Mar.	3	20·6	XVII	—	—	0·74	0·71	0·81	1·0
1 May	4	14·3	XX	—	—	0·98	0·85	0·70	1·0
21 May	4	14·3	XXI	1·30	—	—	—	—	1·0
Mean ultimate air ratio				1·12	0·83	0·74	0·78	0·75	1·0

From these nitrogen data we may make a selection of the cases that will correspond, in ontogenetic stage, with our low O_2 apples. As, in our present work, we have one long sequence (for 5 % O_2) ranging from apple IX to apple XX, we desire to bring in such nitrogen cases as will cover a similar range. We select VII, XI and XXI*a* and *b*, omitting XIX, as its record was based on short and very irregular data. In fig. 2 we have plotted the drift of NR/ALR ratios for the four selected nitrogen cases. For our table 5 we need the values at 50 hr. in nitrogen as the best correspondence with the subadjusted values in low oxygen concentration. For XXI we take a value of 1·30 intermediate between the 50 hr. values of XXI*a* and *b*, as this corresponds best with the time relations of the ratios from the comparison apple XX. For apple XI in nitrogen we have to carry out a short extrapolation as is evident from fig. 2, which indicates 1·06 for the 50 hr. ratio. For an earliest case we have no choice but to make use of VII, so we may take 1·0 as the 50 hr. ratio.

Having added these nitrogen ratios to those for low oxygen and completed the set with the standard value of 1·0 for air, we find that contemplation of table 1 produces rather a confused effect. There stands out, however, a marked tendency to produce the lowest ratios round about 5 % O_2 and the highest values at each end of the oxygen scale. On the bottom line of the table the *mean* value for each column is set out; and this makes CO_2 production in 5 % appear as about three-quarters of that in air. In fig. 3*a* we have plotted all the tabulated values over an oxygen concentration axis of O_2 supply, and have drawn the curve of mean values through the points. The conclusion that a minimum exists cannot be evaded, though the high ratio of XX in 5 % O_2 weakens the graphic impression of this feature.

The minimum in the distribution of these adjusted ratios is of course only another graphic expression of the change in the direction of movement of the ultimate air ratios, between the simple and the compound transition, in fig. 1, located at the critical transition in critical O_2 supply.

Obviously this collection of air ratios requires careful sorting and arrangement if a clearly defined pattern is to be brought out and substantiated. The ideal would be that each experiment, as identified by its chronological roman numeral, should contain a fair range of different O_2 supplies so that a large set of single curves could be drawn and compared. Only XVII and XX are rich in significant values, so the experiments must be combined into groups, each being just large enough to give values for a curve from nitrogen through to air. It is clear that the high chronological numbers give the high ratios at the nitrogen end of the scale, so that a stratified separation of curves on an ontogenetic chronological basis promised to be manageable.

There are four indexes of ontogenetic state to which attention may be paid, namely, (1) the numeration sequence of the apples, (2) the month of withdrawal from storage for experimentation, (3) the calculated initial air-pitch at 22° C. on withdrawal, and (4) the type of transitional drift for 5 % O_2 supply as lettered *c, d, e, f* on p. 108.

It was not possible to break the collection up into more than four groups and the numerical components of these are set out in the table which makes part of fig. 3*b*.

As suitable for curve 4, fig. 3*b* and *c*, the ratios for XX seemed remote from the others and very much higher in 5 and 7 % O_2. A nitrogen value was needed to complete this set, and the mean of XXI*a* and *b* offered itself as an appropriate value. It may be mentioned in support of this value of 1·30 that it is actually also the mean of the 50 hr. nitrogen ratios for XIX, XXI*a, b* and XXIII. It is very important to have a well-established nitrogen ratio for the start of the highest curve. For curve 3, fig. 3*b* and *c*, XVII will have to be a conspicuous feature. Apples XVI and XIV suggest themselves for this stage of senescence, and this is held to carry XV into the group. There was unfortunately no apple between XI and XIX, which had been examined in nitrogen. An imaginary ratio of 1·11 is inserted within brackets in table 1, the choice of this value being guided by the general relations of the other curves at the nitrogen end. Four apples now remained representing the earlier ontogenetic stages of metabolic development, and we segregated them into two separate curves, as the ratios in 5 %

indicate so much divergence. Curve 2, fig. 3b and c, contains apples XI and XIII, so that it has real values in 0, 3 and 5% O_2, while curve 1 is made up of VII and IX, presenting only one other datum beside the low 5% ratio.

We have next to draw the best empiric curve we can right through from air to zero O_2 for each of these four groups. The first aid we need is a general idea of the

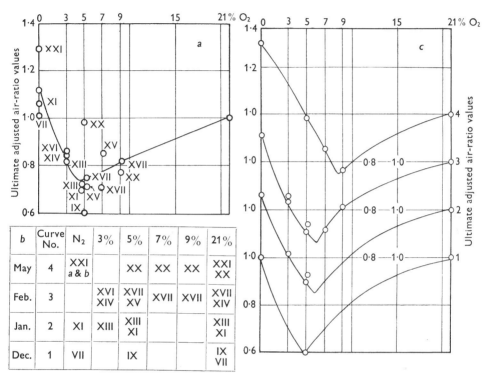

Fig. 3

Note to fig. 3.* All the ultimate adjusted air ratios (TR/ALR concurrent at hour 50) of table 1 are plotted in fig. 3a over an oxygen concentration axis of O_2 supply. The roman numerals indicate the respective apples. The curve is drawn through the mean values.

The collection of air ratios of fig. 3a is separated into four groups in fig. 3b according to the indexes of ontogenetic state set out from p. 118 onwards.

The four groups of air ratios of fig. 3b are plotted in fig. 3c, each group with a different ordinate setting. The method of drawing curves 1, 2, 3 and 4 is described on p. 120. Fig. 3c illustrates the influence of O_2 supply and ontogenetic change on the relative rates of CO_2 production in the adjusted states.

slope† of the curve as it bridges over the large gap between 9 and 21% O_2. We have only one value beyond 21% namely the air ratio of 1·4 at 100% O_2 supply, and we have drawn all our curves at the air end as if they would slope through 21% to attain a value of 1·4 over 100% O_2. It must be stated that we have no picture of the effect of ontogenetic drift upon air ratios above air.

* A legend has been added since none could be found.

† These graphs of varying CO_2 production over an axis of O_2 supply we speak of as curves of rising or falling CO_2 rates in contrast with the terminology for graphs of CO_2 production over a time axis for which we speak of drifts of inclining or declining CO_2 production.

With this common start at the high oxygen end we settled down to draw four separate curves, one for each group, with the bias of giving them similar forms where there were only few established points for guidance. In fig. 3*c*, for clearness, the four curves are drawn as separated by short distances in the vertical direction, each having its individual ordinate values repeated against it. The points of curve 3 clearly indicate a sharply defined minimum, and such sharpness has been given, by free-hand drawing, to the other three curves also. The crux of the drafting was to approach the minima from the two sides by curves extended until they intersected. We may indicate the steps that we took to construct this sequence of four curves.

With curve 4 the descending line from the nitrogen side, after passing through three points gives the line for the minimum very narrowly from one approach. On the other side, if we start out from the ratio 1·0 with the slope indicated by the 100 % O_2 ratio of 1·4 and have to pass through the ratio of 0·77 in 9 % O_2, then the intersection of lines defines the minimum very closely. The same applies to curve 3. Having drawn these two curves one notes that the minimum has drifted to being over a lower concentration of oxygen. Looking in advance at curve 1 we see such a low ratio, 0·60, that there is no reason to regard this as any other than the actual minimum value. If so it harmonizes with the view that the minima continue to be displaced nearer to nitrogen as they become lower on the ordinate axis. On the strength of this, the minimum of curve 2 is put intermediately between 3 and 1 and connected up with air by a curve similar to that of 3. Finally, the sketchy but important elements of curve 1 are connected together by lines resembling those of curve 2.

Fig. 3*c* now provides us with material for defining some elements of the pattern of respiratory metabolism for this tissue. We note two principles: one concerns the effect of external O_2 supply as shown by the form of all four curves in relation to low and very low O_2 supply, and dictates that somewhere or other there shall be a well-marked minimum, while in nitrogen we get higher values than in air. So much any one curve would have taught us, but the ranging ontogenetic set teaches further that the two co-ordinates defining the minima shift with ontogeny so that neither the air-ratio value on the ordinate axis nor the O_2 supply on the oxygen axis remains unmoved. Further than this it may be considered to be established, though the drafting has contributed to the apparent sharpness of the demonstration, that the direction of drift of the location of the minimum with rising ontogeny is oblique, moving both to higher air ratios and to higher O_2 supplies. This movement is defined in extent in table 2, where the co-ordinates of the minima of the four curves are set out.

The metabolic interpretation of these patterns will be taken up in Part IV.*

Leaving aside this consideration of our analysis of the pattern of respiratory metabolism we may conclude this section by building these four curves into one figure to represent the combined effect of O_2 supply and ontogeny on the relative respiration rate of apple tissue. Fig. 4*a* provides a full close to a section of this work, as it is a self-supported result independent of all transitional states which

---

* Part IV must refer to Papers VII and VIII.

have figured so largely in constructing it. As an experimental result, it will be remembered that the values in this figure were measured at a point of time just 50 hr. after the effective cause (a change in external O_2 supply) had been applied. To display the extent to which this pattern would be distorted by taking similar

Table 2. *Minimal air-ratio values, with their location on the axis of O_2 concentration*

Curve	Air ratio TR/ALR	% O_2
1	0·60	5·0
2	0·65	6·0
3	0·67	6·3
4	0·75	8·7

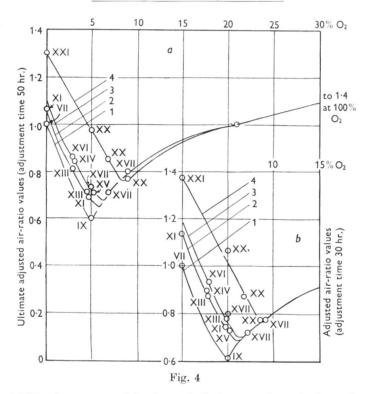

Fig. 4

Note to fig. 4.* The four curves of fig. 3c are built into one figure in fig. 4a by plotting to a common ordinate scale.

The construction of fig. 4b is similar but the adjustment time for the ratios was 30 hour instead of the 50 hours for the values of fig. 4a.

measurements before adjustment to the new state was complete, we have examined the tables 1, 2 and 3 of Paper IV and collected values for a period 30 hr. after the change instead of 50 hr. These are not tabulated in this paper, but the inset fig. 4b has been constructed from them, presenting here only the central part of the

* A legend has been added since none could be found.

system for comparison with the adopted system. In fig. 4*b*, curve 1 remains unaltered because, as we noted above, the early senescent stage may achieve 'adjustment' to changed oxygen within 30 hr. Curves 2, 3 and 4 are, however, all shifted upward by the use of values which are not yet fully adjusted. It is satisfactory that the general appearance of the pattern is so little altered though the numerical relations of parts have been distorted to a certain extent.

We may present fig. 4*a* as the first comprehensive schema, that has been put forward, for a higher plant, for the effects of oxygen supply upon the full range of respiratory metabolism between the points of the air supply and the zero supply in nitrogen.

The metabolic interpretation of these features in terms of carbon and oxygen will be taken up in Paper VI, and in Papers VII and VIII the analysis of TR into its components OR and NR will be studied quantitatively.

4. The Analytic Value of the Initial Air Ratios in very Low O_2 Supply

In contrast with the previous section on ultimate air ratios which may be described as self-contained and definitive about rates of respiration in low O_2, the present section starts a line of analysis of the compound nature of respiratory metabolism which at once leads on to other problems. The treatment of the initial air ratios here will therefore be of a preliminary nature just indicating the metabolic problems that they present, which will be fully expounded in Paper VI, § 3.

The conception of a significant 'initial air ratio' was introduced in the general presentation of the course of the 50 hr. transitions that follow change of O_2 supply (see p. 88). It was there noted that transitions to very low O_2 supply exhibit a feature which is absent from the simple class which we have termed empirically an 'early peak' of CO_2 production.

These early peaks of TR were obvious enough to call for searching examination. Had they been passed over as mere 'preliminary states' to the settling down of respiration, a special opportunity for analytic investigation would have been missed.

Our manipulation of this initial air ratio is an attempt to abstract from the compounded slow transition one component of metabolic change that acts very quickly so as to evaluate it in isolation from the total slow adjustment requiring 50 hr.

In Paper VI there will be a full metabolic exposition of the form and dimensions of these transitional features, but we may here state in a few words the significance of the early peak of TR values from which this analysis starts. When the tissues are suddenly changed to *very* low oxygen an oxidation deficit is conditioned in the tissues and the observed rise of TR is the response of respiratory metabolism to this condition. This response consists in the appearance of fermentation in the tissues where before all had been aerobic respiration.

The first essential was to obtain numerical values giving expression to these early peaks which showed a large range of magnitudes; from large ones, on transitions to nitrogen, down to hardly perceptible rises in O_2 supplies just below the critical supply. In supplies of O_2 which are described merely as low O_2 there is no sign of the early peak, the transition having a simple instead of a compound form. To obtain accurate relative evaluation of these peak effects we have in all cases extrapolated back to zero hour the declining series of values that follows the summit of this peak in TR production. We thus arrive at an 'initial maximum' of TR which we justify as an approximate measure of the response of the respiratory metabolism, could the oxidation deficit have been created at the instant of external change and simultaneously sprung to its full external response in TR units. Actually, in any such change, taking place in tissues of considerable diffusive resistance, the readjustment of gaseous exchange requires about 9 hr. for completion, and during this period the rates of metabolic production of CO_2 are obscured by physical processes of solution and diffusion. These purified initial maximal TR values will then serve to provide rates of TR relative to air as unity by presenting them as initial air ratios of the standard form TR/ALR concurrent.

Our experimentation in very low O_2 was not planned to provide a large collection of subcritical initial air ratios, and we can only find ten fairly well-established examples to bring into this preliminary survey. Among these, the three for zero % O_2—VII = 1·50, XXI a = 1·52 and XXI b = 1·51—were borrowed from the earlier paper on nitrogen (Paper II, p. 42). Three others are provided by the experiments with 3 % O_2—XIII = 1·22, XIV = 1·34, XVI = 1·25 and four others may be put together for 5 %—IX = 1·0, XVII a = 1·20 and XX a = 1·31; these values are entered in tables 1 or 2 of Paper IV, while there may be added a rather uncertain value of 1·1 which was suggested for the group XI, XIII and XV in the discussion in § 2 (p. 113), as being more strongly indicated than the value of 1·0 allotted to these in table 1, Paper IV.

Though these ten examples are not enough for a critical survey of the initial function they should suffice to indicate the general pattern of its control. The obvious step to take is to draw a graphic figure for the initial air ratios on the same lines as fig. 4 a for the ultimate air ratios on p. 121.

This is presented in fig. 5, and we may point out first how the two systems differ in relation to the axis of O_2 supply and afterwards glance at the displacement due to senescence.

The ten values, which are clearly marked in the figure, show that there is the same broad movement of values in relation to low supplies of oxygen, in that the highest values occur in nitrogen while lower ones are proper to 3 % O_2 and still lower to 5 % O_2. The initial ratios never fall below 1·0, and in all the higher O_2 supplies up to air and beyond, the initial ratio will remain exactly at 1·00, which establishes a difference from the ultimate ratios.

We record but a single value for all three nitrogen initials, although the examples were as far apart in ontogeny as VII and XXI, so that a further divergence from the behaviour of the ultimate ratios is established. On the other hand, for the range of 5 % O_2 initials we find that the rising order of the individual cases is that of senescence, in close agreement with the behaviour of the ultimate air ratios.

Granting that the 3 % initial ratios are a little mixed, as they were also with their ultimate ratios, we may formulate the preliminary proposition that the *initial* ratios in nitrogen are not affected by ontogeny, while those for 3 and 5 % are affected to a rising extent. This effect is in the sense of moving the location of any selected ordinate value further along the O_2 supply axis as senescence increases. There is no narrowly located minimum ratio in this set, but the minimal value of 1·0 has a wide distribution. The locus of the low O_2 supply at which the value of 1·0 starts is seen to shift, with rise of senescence, towards air just as with the diagram of the ultimate air ratios.

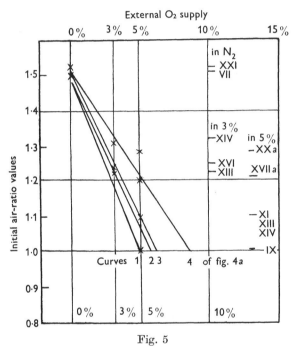

Fig. 5

Note to fig. 5. Values of initial air ratios (obtained by initial extrapolation for all values above 1·0). The × × indicate the values inserted from experimental observations, there being 3 for nitrogen, 3 for 3 %, 6 for 5 % (the apple numbers for the individual cases are added in groups to the right of the figure) on the appropriate ordinate level.

As a background for these values are drawn the four curves of the ontogenetic groups 1–4 recognized in fig. 4 giving four steeply falling lines from a common origin at 1·5, then diverging with senescence to cut the ordinate minimal initial ratio of 1·0 at the four loci of ext. O_2 supply recognized as critical in table 2, p. 121. For all higher external O_2 supplies these curves would continue indefinitely at the 1·00 level.

It will be seen that a bundle of four sloping straight lines has been drawn on the figure to meet the abscissa axis at the four values of O_2 supply which were adjudged to be the critical locations for the four ontogenetic curves in fig. 4a. All these start from the same pitch of 1·5 in nitrogen and all end with the value of 1·0 at critical concentration ranging from 5·0 to 8·7 %. The pattern that seems to emerge from this figure is that each line stands for an ontogenetic stage on which the initial air ratios of the O_2 supplies between zero and the critical supply would be strung out. In addition, it appears that the slopes of these lines get less steep with rising senescence.

Passing into the supercritical region we can only expect to get the initial air ratio at 1·00, which is another way of saying there is no early peak or rise of TR. Behind this absence of surface effect lies the fact that the effective cause of the rise is absent; the defined shortage of O_2 has not been developed by the minor lowering of supply. and there is still sufficient to oxidize all the active carbon at the rate it is being produced at the initial point of time. These metabolic indications find full development in Paper VI.

VI

THE METABOLIC INTERPRETATION OF THE
OBSERVED RATES OF CARBON DIOXIDE PRODUCTION*

CONTENTS

INTRODUCTION

Our object in this Paper VI is to put forward specific metabolic interpretations of the major variations of the rate of CO_2 production that we have found to follow change of external oxygen supply to respiring tissues.

We have attempted to segregate and consolidate this field of metabolism by uniting together the phenomena observed in nitrogen with those observed in all ranges of oxygen supply, under the heading of *Respiratory Metabolism*, aerobic or anaerobic. All types of this energy-releasing metabolism in the higher plants involve interactions of the elements carbon and oxygen in one way or another.

The plant's relation to these two elements are widely contrasted. Life in a medium containing 21% O_2 must have been the least variable of all environmental features throughout recent evolution. Yet at a touch the investigator can alter the normal oxygen supply through a range from zero percentage to pure oxygen at 20 atmospheres pressure. A study of the effects of such changes should therefore be instructive with regard to one type of biological mechanism.

The supply of carbon compounds for respiration is, in strong antithesis, kept under the plant's intrinsic control and is subject to developmental and other changes which the investigator is powerless to arrest. Extreme vigilance and a wide range of experimentation is called for to get standardized results upon this shifting carbon background.

As an introduction to our conception of respiratory metabolism as a carbon sequence of the production and consumption of carbon metabolites we set out in four lines a schema of symbols that we propose to make use of:

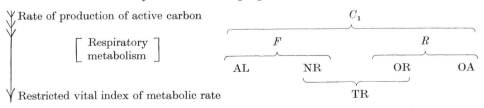

* This paper has not been published previously. No authors were named on the typescript. The latter was dated 'August, 1937' in Blackman's writing.

Respiratory metabolism, in our segregated sense, is held to be initiated when activated carbon compounds, proper to carry on the later stages, are formed from inactive carbon sources which we shall leave undefined at present. In our abbreviated symbolism this initial stage is termed the production of active carbon (C_1) from inactive carbon (C). The rate at which this production proceeds in any tissue at any state is labelled the 'Carbon grade' of the tissue. For the next stage, after C_1, there are alternative fates according to O_2 supply, with the class labels fermentation (F) and aerobic respiration (R). Each of these classes has a gaseous end product NR or OR in association with a 'tissue product' (AL) or (OA). The sum of the gaseous products produced and escaping from the tissues can be accurately measured for indefinite periods during the life of the tissue, and this provides us with an invaluable vital index of respiratory rate TR.

For the purposes of our particular study of an aspect of plant metabolism we have simplified respiration to the puerile formulation of $C_1 + 2O_1 = OR$. This may appear very neglectful of the wealth of biochemical investigation now carried on in endeavours to individualize all the molecular forms involved in respiratory mechanism. Perhaps the opposition will appear more like co-operation, to a common end, if we stress that our objective is to generalize the metabolism into classes of reactions and the biological control of their rates. This we take to be the proper objective of physiology as segregated from biochemistry which seeks to individualize the reactants and the products of intermediate metabolism.

The real problems of protoplasmic vitality should yield something to a purely kinetic study *in vivo* of metabolic activation, rates and controls which cannot be delayed by waiting for identification of the precise metabolic reactants involved.

All the symbols in this schema, just set out, carry a quantitative significance based numerically on their numbers of units of contained carbon, so that metabolic sequences and equivalences can be drawn up in quantitative terms. The totality of drift through the metabolism, as here formulated, must be equal at the initial and final stages, so that the sum of the four products should in carbon be equivalent to the sum of C_1. As, however, the tissue products cannot be directly measured during life it follows that the vital index TR is a 'restricted' quantitative index, speaking for only part of the total carbon involved. If, however, it is established that constant ratios hold between AL/NR and also between OA/OR so that $R = x \times OR$ and $F = y \times NR$, then it will become possible to derive from TR an important measure of the rate of the full respiratory metabolism equivalent to the totality of C_1.

In the present paper we limit ourselves to inquiries about the amount of the restricted metabolism, i.e. that part of C_1 which finds its way out of the tissues as TR. For our immediate inquiry which is about the effects of O_2 supply upon all this respiratory metabolic rate, this restricted index appears to be quite satisfactory, because our present data are all relative ones. We do not deal with absolute rates of TR, but with special 'air ratios' of the form TR/ALR, so that all rates are relative to rate in air taken as unity. Only if the make-up of R as between OA and OR is different in different oxygen mixtures shall we have to hesitate about taking the restricted index TR as an accurate indicator of the behaviour of the full respiratory metabolism C_1.

We have seen in our long experiments at constant temperature on individual tissues that 'carbon grade', though steadily changing owing to internal controls, provides the biologically continuous element in the determination of respiratory rate, slowly unrolling on our records a line of CO_2 production which embodies the effects of temperature, starvation and ontogenetic drift. We may view carbon as constituting the 'warp' of the pattern of controlled respiratory metabolism, while the extrinsic temporary variations of O_2 supply due to the experimenter constitute 'the woof'. The carbon warp provides the labile plant element of the pattern which can be displaced by change of O_2 supply, but only to a limited extent dependent on the plant's metabolic mechanism.

1. The Production Rate of Active Carbon and the Dual Interaction of Oxygen in Control of CO_2 Production

We propose to treat the production rate of C_1 as a metabolic entity. The evidence that has led us to adopt this view cannot be detailed till other aspects of our respiratory survey have been paraded. At present we will introduce this conception by making four assertions about its characterizing features, when the function is surveyed through the whole range of external O_2 supply. These features are supposed to hold for the carbon grade at all stages of starvation and ontogeny, with relation to O_2 supply. For any such state:

(i) There is a minimal rate of C_1 which we shall call the 'deoxygenate grade', which prevails not only in complete absence of oxygen, but also in conditions of inadequate O_2 supply such as will shortly be defined as '*very* low O_2 supply'.

(ii) When the O_2 supply rises above these inadequate values, though it may still be classed as 'low O_2 supply', then this carbon grade rises, undergoing what we term 'oxygen augmentation'.

(iii) The value of external O_2 supply which marks the boundary value between the deoxygenate and the O_2-augmented rates of the carbon grade is to be known as the *critical* external O_2 supply.

(iv) With rising supercritical external O_2 supply the augmentation of carbon grade gets higher and higher up to a pure O_2 supply, and even beyond. Ultimately, with oxygen at several atmospheres pressure, the carbon grade declines and may be forced down to zero, with irreversible cessation of all respiratory metabolism.

The relations of carbon grade for the early storage stage of apple tissue are set out in fig. 1 A. The ordinate values are *relative* rates, that in air being given the standard value of 1·0. In relation to this the deoxygenate minimum for this tissue would be 0·6. This value holds from zero O_2 to the critical supply here identified arbitrarily with 5 % O_2 external supply on the abscissa axis. Above this we see the form resulting from the rising degrees of O_2 augmentation. This rise would continue to an ordinate value of about 1·4 in pure oxygen, but the record is here broken off at 21 % O_2 supply.

The dual effects of oxidation and oxygen-augmentation. Here we shall deal first with the oxidation of the active carbon, C_1, by interaction with active oxygen, O_1, and then take up the augmentation of C_1 attributed to oxygen in the previous paragraph.

The external O_2-supply to the respiration chamber that the experimenter provides sets up a diffusion potential for the entry of O_2 into the tissues, functioning in proportion to lowness of O_2 pressure maintained in the tissues by the biochemical consumption of O within them. If the catalyst activating O_1 is in excess, through the range of oxygen conditions that we are to consider, then the *potential* production of O_1 will rise in direct proportion to the external O_2 supply. In fig. 1 B a rising straight-line slope of this function is drawn from zero O_2 to the supply in air. This slope might be given an ordinate scale of oxygen units of consumption relative to the consumption of O_2 in air, taken as unity, but as our present outlook is based on carbon units we can substitute one C_1 unit as equivalent to a pair of O_1 units, in oxidation to CO_2 and so pass from the O_1 scale on the right to the C_1 scale on the left of this graph, and make comparisons with fig. 1 A for C_1. As there can be no consumption of O_1 units apart from production of activated carbon a super-position of the C_1 curve (*a*) upon the potential O_1 curve (*b*) in this figure would give us a clue to the actual consumption, by union of O_1 and C_1 in the different external O_2 supplies, by consideration of their respective excess or deficit. We may first make the point that, as O_1 rises more steeply than ever does C_1, there must be an intersection at some locus, which we assert to be the critical O_2 supply of this set of figures

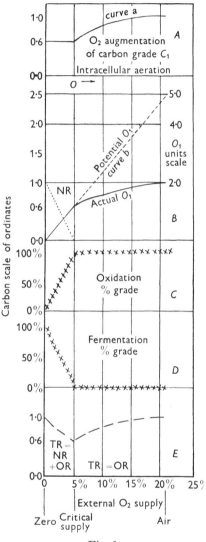

Fig. 1

generally and which locates the equivalence of rates of actual consumption of O_1 and C_1 in producing oxidative CO_2, both having here the relative rate of 0·6. It is clear then that in the regions of O_2 supply above and below the critical O_2 supply the oxygen conditions in the tissues will be in strong contrast. In very low O_2 the O_1 supply will be in deficit compared with the deoxygenate carbon grade of 0·6; and O_2 will therefore be completely consumed as it enters, the actual consumption keeping identical with the potential, up to the critical point. To the fate of the excess production of C_1 in very low O_2 we will return shortly after considering the actual O_1 consumption in the supercritical region. Here O_1 rises more steeply from the 0·6 value than does the C_1 curve, taken over from the top figure. Here there is potential O_1 in excess, and therefore the whole of the active carbon produced will be oxidized to CO_2. Since the carbon can no longer consume the whole O_2 that the external O_2 supply could cause to enter for activation, then aeration within the cells will begin just at the critical supply, and at each rise of supply the 'residual O_2' will maintain a higher concentration of free O_2 within the cells.

It is to this intracellular O_2 condition that we attribute the O_2 augmentation of C_1 production that we have represented in fig. 1 A.

The mechanism of this augmentation of carbon grade by residual free O_2 is held not to involve any appreciable or continuous consumption of oxygen, but to be due to the establishment of equilibrated states between the partial pressure of O_2 and the balance of oxidized versus reduced states of some constituent of the enzyme complex that controls the rate of production of active carbon from inactive carbon sources. The widening gap between curves a and b in the figure as O_2 supply rises, indicates the rapid increase in the maintained intracellular aeration that should follow the experimental progress through the low O_2 supplies towards that proper to air. Experimental studies of oxygen conditions, biochemical and biophysical, may provide concentration values for this condition; so a blank scale is left at the upper part of fig. 1 B, marking only the zero value as coincident with the critical external O_2 supply.

We must now turn back to the subcritical region and examine the oxygen condition provided there in regard to its effect upon the C_1 production rate which is clearly in excess of O_1. As fig. 1 A shows, the rate of C_1 is asserted to keep its constant minimal value throughout the region. In nitrogen the whole value of C_1 undergoes fermentation and gives rise to the full maximal production rate of NR as the gaseous product. Here then we meet a kind of standard condition which contrasts markedly with the standard condition of AIR which we have adopted for present guidance, and note that use will be made of it in subsequent analyses. In fig. 1 B this rate of production of NR is shown by a falling line of fine dots, with its maximal value in nitrogen as 1·0 relative to the CO_2 production in air. [Such an equality of values is merely an incidental character of the ontogenetic state of the particular tissue adopted for this exposition.] Moving from zero along the abscissa axis to give a trace of external O_2 supply, a small fraction of the deoxygenate C_1 rate will be oxidized to OR by the total consumption of the O_1. As a result the escaping TR will largely be NR, with a trace of OR. A larger O_2 supply shifts the fraction of OR to something higher, with decrease of NR. So with rising O_2 we note the progressive substitution of OR for NR in ways that will be expounded quantitatively in a later paper.

When we arrive at the critical O_2 supply and O_1 has risen to equivalence with C_1 there is no longer fermentation, for C_1 is just completely oxidized. Hereafter there is excess of oxidative power and the carbon continues completely oxidized.

We may express this progress of oxidation, at each stage, as the percentage grade of available C_1 which is oxidized by O_1 to CO_2 and fig. 1 C figures this oxidation grade rising from 0 to 100 % in the critical O_2 supply and then remaining at 100 % for all higher values of C_1. To complete the graphic picture of these relations we add fig. 1 D, which presents the 'fermentation grade' which is simply the linked opposite movement from 100 % in nitrogen to 0 in critical O_2 supply.

2. The Limiting Relations of Carbon and Oxygen Controls in Adjusted States of the Tissues in Low O_2 Supply

In fig. 4*a* of Paper V we have already presented the facts of CO_2 production in a range of lowered O_2 supplies, as observed after the new supply has been carried on for 50 hr. so that adjustment to the altered supply is complete. The numerical values used in that figure are the values of the TR/ALR ratio which gives the relative TR values when that in air is taken as unity. There are four curves of similar forms in the figure as the whole range of ontogenetic metabolic drift involves a shift of cardinal points. We may take over curve 1, that of the earliest storage stage, and transfer it to our present group as curve *E* in Fig. 1. In this company it stands for the unanalysed observed curve of total CO_2 production through the studied range of low O_2 supplies, and we have to show that this synthetic curve of direct observations can be satisfactorily interpreted by the metabolic principles we have been expounding.

This curve of TR shows a marked break in form where the CO_2 production reaches its minimum, and we are presented with the remarkable side feature that a rise or a fall in O_2 supply alike raise the CO_2; which proclaims at sight some duality of control. This minimum locates the critical O_2 supply, which has shown itself for the curves of the carbon supply and the O_2 supply. Looking back at what we have just asserted about the various metabolic aspects, we may give a summary statement of the controls and the limiting factors on either side of the critical O_2 supply.

On the subcritical side, TR falls rapidly from its value of 1·0 in zero O_2, declining step by step as the O_2 supply rises, until the critical supply. This is an arresting fact, since we have seen that O_1 potential is rising in this region in proportion to O_2 supply and with it rises the OR arising from oxidation. Here then O_1 is the limiting factor to oxidation and C_1 is in excess until, at the critical O_2 supply, O_1 rises to equivalence with C_1. From zero O_2 to this locus C_1 is held to have kept its minimal deoxygenate value all along, so that the *full* respiratory metabolism has been maintained at a uniform rate (see the curve for the carbon factor, fig. 1*A*). In nitrogen the whole of this carbon was undergoing fermentation, and the principles of its gradual substitution by OR have already been set out.

One of the striking facts about the observed CO_2 with the apple tissue is that NR in nitrogen is larger than OR at the critical O_2 supply, though the same total of respiratory metabolism of carbon is involved throughout the subcritical region. The verbal interpretation of this which will be developed later is that NR is a larger fractional index of F than is OR of R, so that in the expressions (ii) and (iii) on p. 127 x has a higher value than y.

We now pass through the critical O_2 supply which is defined as the only situation in which O_1 is equivalent to C_1, and characterize the controls and limiting factor in the supercritical region. Here the TR form shows a rising curve, with rise of external O_2, which is identical with the curve put forward for C_1. It falls below the straight-line form of the rise of oxidation potential O_1. In this region, C_1 is the limiting factor, O_1 is in excess, the observed TR is 100% OR and there is no fermentation. The control of the rate of C_1 in this region is asserted to be vested in the intracellular grade of aeration, working by the mechanism of O_2 augmentation as already described.

Above the locus of the critical O_2 supply, C_1 falls behind O_1 and becomes now the limiting factor to OR production. Though C_1 undergoes O_2 augmentation as the O_2 supply rises, it never catches up with O_1 but becomes always further and further in retard.

That increased oxygen supply increases the rate of respiration is a generally accepted principle for plant tissues, but it is asserted here that in the biological regions at and just below the normal atmospheric O_2 supply this result is produced by increase of carbon fuel supply for oxidation, not by increase of oxidation directly due to rise of available O_1. We hold that increase of respiration by the action of this latter principle is only at work in the subcritical region of very low O_2 supply. It is only here that carbon is in excess, while in normal biological regions carbon is limiting and O_1 is in excess.

3. The two Types of Action of Oxygen upon Carbon and their Differential Effects upon Transitional Drift Forms

The observed TR values for which we have sought metabolic interpretation so far have been those obtained as fully adjusted air ratios after 50 hr. in any given low O_2 supply. We have now to turn to the drifts of TR values during the 50 hr. between air and adjustment, which precedes the adjusted values.

These drifts show some remarkable and unusual features, and it has taken a great deal of thought to build up one metabolic picture which will interpret the range of these drifts as well as the adjusted TR values which result from them.

In fig. 1 of Paper V we have pictured a model family of such drifts divided into two classes, the simple form and the compound form with the 'critical' boundary form between them. The essential features of these transitional forms concentrate about the 'initial TR values' and the 'ultimate TR values' and the way they differ according to the lowness of the O_2 supply.

We may glance first at the ultimate TR values (or air ratios) as they are identical with the 'adjusted' values just dealt with as fig. 1 E, and the essentials of their metabolic interpretation have already been set out. Throughout the simple family, which inhabits the supercritical region, active O_1 is always in excess, and these drifts express the downward declines of C_1 ending in the set of ultimate pitches which represent the different grades of O_2 augmentation of C_1, the critical lowest being the minimal deoxygenated rate of C_1 production. Turning to the family of compound drift-forms we find that the lower the O_2 supply, the higher lies the ultimate TR value, the maximal being found in nitrogen. Here then superposed on the minimal C_1 we have rising grades of fermentation substituted for respiration, both combining to cause the ultimate output rates of TR (see p. 126). To the C_1 component which lowers the TR production, in this region, we have added the F component as the negative aspect of oxidation and this raises TR, overpowering the effect of the C_1 component in this set of compound forms. We may now turn to the initial TR behaviour through this family of drift forms. The simple forms start on the same sort of drift that they continue to the end, but the compound drifts all exhibit an initial rising effect leading to an early peak. Here then we have a special feature only found in the transitions to 'very low O_2 supply' and appearing as a set of initial rises of different pitches, being highest in nitrogen. Clearly we have thus evidence of duality of oxygen effect, and as the slow decline is attributed to C_1 and lowered O_2 augmentation, the rapid rise must be attributed to the increased fermentation component as the O_2 supply is reduced towards zero.

Nothing that we have asserted in our survey of respiratory metabolism, and the duality of the oxygen effects of oxidation as contrasted with O_2 augmentation, serves to forecast this special effect of an early peak of TR. We have to make a new assertion which concerns the *differential rate of adjustment* of these effects on change to a low O_2 supply. These dual effects of oxygen start off together, but change of oxidation (and inevitably also of fermentation) grade comes to full effect in a few hours while adjustment to a new O_2 augmentation grade may require 50 hr.

We conclude that the slowness of transitions on change of O_2 supply which so much hampered us in the study of a range of adjusted states is not a characteristic of the changed oxidation grade involved but is solely an attribute of the other oxygen effect, the O_2 augmentation grade of active carbon production. This differential rate of adjustment for the two oxygen effects has, however, the invaluable advantage that it enables us to demonstrate the dual oxygen effects separated out in time on the transitional records.

We may gain analytic insight into this compounding of dual oxygen effects if we consider a few possible variations from the make-up here revealed. The two components affecting TR may be styled for brevity the 'decline of C' and the 'substitution of F' for R, giving a production of NR. We may begin with constructing hypothetical drift records in which the two alleged component effects take place quite separated in time, one after the other. We will omit any display of difference of rate of adjustment for the two components in this first analysis.

Fig. 2 displays the manipulation of the two components in one graph, putting the transitional 'decline of C_1' before the substitution of F and the other reversing this order. Each form starts with the same antecedent pitch in air and ends with a fourth period of identical completed adjustment to the very low O_2 supply of, say, 2·5 % (drawn with heavy line) as the central feature, while associated with this (in dotted lines) are the adjustments to the critical O_2 and to nitrogen. The upper record assumes that in the first half-transition C_1 declines to the minimal deoxygenate level which is common to all *very* low O_2 supplies, while, for the time, the oxidation potential is not altered; therefore this C_1 will appear, wholly oxidized, as a pure but lower OR output.

After this change of C_1 is complete, we picture the oxidation factor starting its transition to the very low O_2. In the record a range of three different 'substitutions' of F is allowed for; the central one (2·5 % O_2) giving a TR slightly below that previously in air, the nitrogen one with complete substitution of F gives TR higher than air while the critical O_2 supply gives the minimal TR as recently discussed.

Turning now to the lower record in which the order of the part transitions is reversed, the first half-transition is refused any change of O_2 augmentation so that C_1' will remain at the high level of air, but substitution of F will take place upon

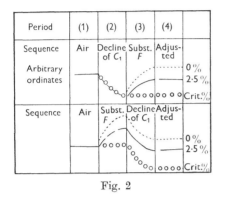

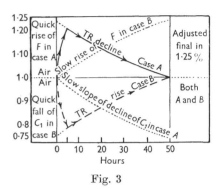

Fig. 2.

Fig. 3.

Fig. 2. Manipulation of the two component changes that make up the transition from air to very low O_2 supply. Between the first period of time in antecedent air and the fourth period in adjusted state to the new very low O_2 are presented two separate periods, one for the component of decline of C_1 and the other for the component rise of F. The order of these components is reversed as between the upper and lower drift records. The ultimate position attained for a given very low O_2 must be identical in relation to the antecedent air whichever time sequence of transition has intervened. For the rise of F component we have a transition to three different pitches of substitution corresponding to the three very low O_2 supplies; while for the decline of C_1 component the transition is the same for all three gases as the same basal minimum value of C_1 will be reached.

Fig. 3. Hypothetical reversal of the rates of adjustment for the two components of transition 'decline of C_1' and 'substitution of F'. A. The apple form of TR drift from air to very low O_2 (say 1·25 % O_2) where substitution of F is allotted 5 hours and decline of C_1, 50 hours. B. The 'anti-apple' drift allotting 5 hours to decline of C_1 and 50 hours to substitution of F. The apple follows the heavy line with an 'early peak' and extrapolated maximum while the anti-apple B in broken line shows an early valley with an extrapolated initial minimum. For case A the curve, in dots, shows the decline of C_1 by itself in the apple; while for case B the curve in dots gives the hypothetical rise of F by itself for the anti-apple.

this high carbon level; curves are drawn out for the same three very low grades of oxidation potential as in the upper record. We thus attain three levels of TR; the highest for nitrogen, the middle for 2·5 % O_2, and the lowest for the critical O_2, which possesses just equivalence of O_1 to oxidize it all. In the second half-transition, the C_1 of the air-level, common to all the three, will decline to the same deoxygenate minimum for all three, and accordingly the three TR's show similar declines, with fermentation grade unchanged, as postulated. The three final levels will drift to the same positions as in the upper record when both have achieved full adjustment.

This imaginary analysis of the transition serves to show how its particular form, as expressed in the observed drift of CO_2 production, is dependent on secondary differential attributes of the dual effects of O_2 supply upon respiratory metabolism.

One other distorted diagram (fig. 3) may be given reversing from actuality the observed differential speeds of the components in adjustment; making both decline of C_1 and substitution of F start at the change of O_2 supply but presenting the C_1 component as adjusting quickly while the F substitution proceeds slowly over 50 hr. This case may be presented as dealing with a very low O_2 supply, say 1·25 % O_2, that might, illustratively, give an adjusted TR value of the same pitch as that of air. The time-scale gives the standard 50 hr. for the full transition, but allows only 5 hr. for the rapid component. The characteristic type for the apple will be recognized in the continuous line drift of TR, rising initially to a peak near 1·25 due to quick partial substitution of F, while the C_1 decline slowly carries the TR down again to 1·0. The other TR drift marked in broken lines shows the supposed quick decline of C_1 bringing TR down to 0·8 at once, from which level TR gradually ascends to 1·0 by the slow substitution of F for R.

In this comparison of transitional drifts we make clear that the normally observed form of drift may be entirely reversed, simply by alteration of the differential adjustment rate of the two component oxygen controls. In extension of this line of thought it might be remarked that if the fermentation grade, raising TR, and the decline of C_1, lowering TR, both proceeded at the same rate through a transition, then the track of TR would continue all through this last figured case at the level 1·00 which it had had in air and is to have again in 1·25 % O_2 when the transition is over. In such a situation a simple TR observed drift would show no arresting disturbance on transition from air to very low O_2, and the line of investigation here pursued might never have been initiated.

VII

EXAMINATION OF THE PHYSIOLOGICAL DETERMINANTS OF CARBON DIOXIDE PRODUCTION BY APPLES IN LOW CONCENTRATIONS OF OXYGEN

CONTENTS

INTRODUCTION

The conclusions of our empirical survey of apple respiration in different concentrations of oxygen were embodied in the graphic form drawn up as fig. 4 of Paper V (p. 121). The essential parts of that figure now appear again as fig. 1, and serve as the starting point of the present paper. Over an oxygen-concentration axis we have four curves of relative rates of CO_2 production, which being brought to coincide at a value of unity at 21 % O_2 diverge widely in lower concentrations while appearing to agree in concentrations higher than air. All four curves have similar forms, presenting higher CO_2 production in nitrogen than in air, but between these values they show a well-marked inflexion, giving minimum values in the region of 5–9 % oxygen.

This set of curves was slowly built up from a multitude of observations by the empirical synthetic procedure described in Papers IV and V. We have now to endeavour to carry out upon the set an analytic procedure which aims at disentangling the functional determinants that lie behind its graphic features.

There are two classes of problems to be faced here: first the interpretation of the special form which is common to all four curves; and secondly, the significance of the drifting divergence which calls for a whole set of such curves, instead of a single curve, to express the CO_2 production relations for apple tissues when these have been drawn from a wide range of ontogenetic developmental states.

As our problems come before us in graphic form, the analysis will be initiated on graphic lines and then proceed to search for the functional activities that lie behind them. We may therefore, usefully, indicate in advance the four stages of

---

* This paper has not been published previously. No authors were named on the typescript. The latter was dated 'Aug. 1935'.

progress in ideal cases. Graphs of the type of fig. 1 are first inspected for (1) *characteristic features*; these are formulated in terms of (2) *graphic determinants*, such as pitch, slope, curvature, inflexion, ratios between defined points and locations upon the abscissa axis. We then seek to interpret this graphic analysis by putting forward (3) *physiological determinants*. These have to be examined for their essential nature, for their grades of independence as variables, and for their ontogenetic variation. Ultimately these are referred to (4) the *metabolic functions* from which they derive.

As we propose to set out in due sequence, on quantitative lines, the interaction of some four metabolic functions, and to examine at least five physiological deter-

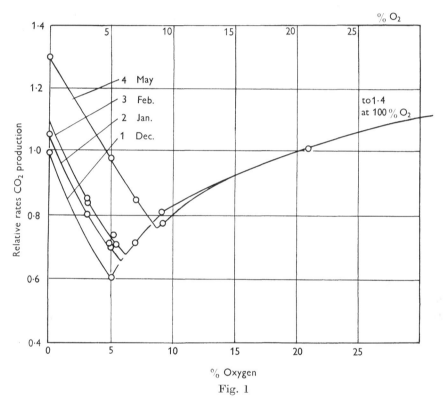

Fig. 1

No legends have been found for the figures of this and the following papers. Because of the full descriptions in the text it has not been thought to be necessary to prepare legends.

minants, we have had to invent a variety of terms and symbols, contrived for brevity and, providing also, we hope, precision of expression. Also we attempt to relate these terms by a number of equations which are gathered together on pp. 141–5.

For the purpose of the quantitative computations which are the essence of our analysis we need to traffic with magnitudes which, though not really of a high grade of accuracy, must be given a pseudo-precision of statement. We therefore claim a good deal of latitude in adopting numerical magnitudes based upon the rather shifting foundations that were alone available for this stage of the survey.

1. The Analytical Interpretation of the Compound Curves of CO_2 Production

(a) *The resolution of* TR *curves into their* OR *and* NR *components: location of the extinction point*, E.P.

The special features of the TR *curve.* The CO_2 production, whose manifestations are displayed in fig. 1, is obviously a mixed phenomenon consisting of anaerobically produced CO_2 NR (nitrogen respiration)—at one end of the oxygen axis and aerobic respiration—OR (oxygen respiration)—at the other end. We adopt for this composite production of CO_2 the term TR (total respiration).

After examining the characteristic features of such a TR curve we shall aim at resolving it into its OR and NR components. We shall thus have two new curves of rate of CO_2 production plotted against the same axis of $[O_2]$ (our symbol for external oxygen concentration) to help in analysis of the situation.

These TR curves present a most unusual graphic feature in that they have a minimum somewhere in their middle region. From this minimum the curve rises on either side towards a peak, giving one maximal value at the high $[O_2]$ end and another, generally higher still, at the zero $[O_2]$ end. Curves of the general rates of functions plotted against an external 'factor' typically show the reverse form, rising to a high peak in the middle region, at the optimum condition, with declining rates on both sides. A minimum in the middle range is most exceptional and provides us with no less than three features of rate of CO_2 production to be dealt with: the two peaks at the ends and the intervening minimal value. As an accessory but remarkable feature we are presented with a sharply marked upward inflexion of the TR curve starting from the TR minimum value. As further features we note that the four TR curves differ from one another in their pitch in low oxygen and also in the locus of the TR minimum.

We have, then, before us a variety of major graphic determinants as well as minor ones relating to slope and curvature.

Seeking to relate these graphic features to physiological determinants we stress that as the only metabolic product we are now concerned with is carbon dioxide there must be a carbon side as well as an oxygen side to the functions at work. The form of graph provided in fig. 1 takes due account of the oxygen side, all the rates being set out over an $[O_2]$ axis, but any effects due to variations on the carbon side of the metabolism are concealed and can only be brought to light by careful analysis. We therefore, as a very important preliminary, register our belief in a latent *carbon function* acting as a *carbon physiological determinant*. This essential part of the mechanism will be laid aside temporarily in favour of the oxygen side of the metabolism.

The effects of oxygen supply to apple tissues are very complex and we shall be compelled to make a long story of their analysis. It will be well therefore to supply the oxygen axis with some physiological signposts in addition to the concentrations of O_2, thus providing a functional scale of general application. These consist of four significant letters along the abscissa axis. They can be used as suffixes to be attached to any rate value of TR, OR or NR to indicate its physiological location along this axis. For these locational suffixes we propose Z and N, E.P., A, M. N and Z indicate the rates in the zero concentration of O_2 (nitrogen); A in 21 % O_2 (air). These two have a defined location on the concentration axis, while the other two are defined by internal states of functioning and accordingly exhibit drift along the O_2 axis from case to case. The suffix E.P. connotes the metabolic extinction point of NR, while M connotes the maximal respiration value in the optimal concentration of oxygen. This lies well outside the region of low oxygen that we have now to study, and for apples is located somewhere about 100 % O_2.

Having sketchily introduced both carbon and oxygen as the functional backgrounds to the metabolic situation, we must start our search for more precisely defined physiological determinants. It is obvious that the major determinant of the form of our TR curve is the fact that it combines the expression of two highly individualized functions which we may conveniently label the *OR function* and the *NR function*. These are two out of our four enrolled functions. While they are highly contrasted in mechanism, they are nevertheless united by several special linkages, the elucidation of which is a fundamental objective of our study.

While the TR manifestation combines both these functions we cannot tell yet whether the resolution of TR will simplify the physiological determinants behind it or complicate them by some special interactions of OR with NR. On broad lines, the distribution of the manifestations of these two functional activities along the $[O_2]$ axis can be simply stated.

In zero external oxygen the OR component must be nil and it will increase with rising $[O_2]$, passing successively from OR_N through the rising values $OR_{E.P.}$, OR_A on to OR_M. Conversely the NR component must be at its maximum in zero $[O_2]$ and decline with rising O_2 to reach its zero value at E.P., as located by the TR minimum.

Below E.P., then, the TR curve is a mixture of OR + NR, and the slope of TR will have no primary significance. Above E.P., on the contrary, the TR curve is judged to be wholly OR, and its functional determinants here must be wholly those of an OR function, often to be spoken of as the (TR→OR) curve.

Our next objective is to interpolate the course of the early part of the OR curve between the values of OR_N at the origin and $OR_{E.P.}$ at the functional break in the TR curve. Having arrived at an acceptable course for this OR component of the observed TR curve we can obtain the NR component by difference.

The interpolated OR *curve for low* $[O_2]$. Could we not collect any criteria of acceptability of curve form for OR, then an infinity of pairs of OR + NR curves might be drawn as possible components of the TR curve in low oxygen.

In the absence of a series of well-attested curves from other plants, a consideration of the nature of the OR function will be used to set some limits to the possible curve forms.

In the simplest possible biological cases of respiration oxidations, the form of rate curve that we encounter may be regarded as compounded of a framework of two straight lines, one a steeply rising upslope from the origin and the other a horizontal line of upper limit, above which further increase of oxygen supply does not carry the rate. The transition of the actual rate curve from the rising line to the horizontal line is not abrupt but shows various degrees of curvilinear transition. According to the length of its course from the origin for which the curve follows the straight-line course of the framework, so we get the appearance of a *rectilinear* track or of a *curvilinear* track, for the early part of the rate curve, which is the part least open to observation.

In biological cases where diffusion of an entering reactant gas plays a dominant part in functional rate, as with CO_2 for photosynthesis or O_2 for oxidation, a high-diffusion resistance will lessen the steepness of the rising slope, as well as modify the curve form at its transition to the upper limit. The outcome of the interaction of the factors may produce rate curves that range from the form of a continuous curve of the hyperbola type to a curve with a localized inflected form. We have to stress here that whether the curve undergoes localized inflexion or continuous, point-to-point inflexion of the curvilinear form, these *inflexions* or changes of slope, are all downward inflexions. There is no precedent for an upward inflexion such as our TR curve shows after the E.P. We cannot pass beyond these simple generalities upon the expected OR form until we have carried out an actual resolution as a first approximation and examined whether major distorting influences have been overlooked.

We may sum up our expectations about the OR curve by the formal procedure of passing from graphic analysis of features on to physiological analysis. We can schedule four graphic features: (1) initial slope of OR curve; approaching (2) an upper limit; and showing intermediately a form which is (3) rectilinear or curvilinear; throughout this course we expect (4) only downward inflexions. For these four features we may suggest a series of physiological determinants. (1) The initial slope is an expression of efficiency of utilization of external oxygen. It is measured by $OR/[O_2]$ giving a coefficient of efficiency. The metabolic function behind this is the oxidative activity of the aerobic respiration system; but co-operating with this is the diffusion facility of the structural organization of the tissues. An increase in diffusion resistance would proportionately decrease the initial slope apart from any lowering of the oxidative mechanism. (2) Changes of the upper limit we may correlate with the carbon function as a control and postpone for the present. (3) The contrast between rectilinear and curvilinear courses is complex in its determination, but we may associate a straight course with high-diffusion resistance relative to oxidative efficiency. (4) Downward inflexions being the normal attribute of OR curves we should attribute any upward inflexion in the course of a curve to a change-over in functional physiological control. When we have satisfied ourselves as to the best course to propose for the interpolated part we shall have four completed OR curves available to add to the set of four TR curves in fig. 1. Each of these OR curves will start at OR_N with the common value of nil and pass to its own $OR_{E.P.}$ value, and then onwards to a common OR_A value of unity passing presently out of our view towards its OR_M value at the optimal $[O_2]$. The divergence that will be found at OR_M between the four curves has not yet been explored.

The curves of the NR *component.* For this curve the only value that is revealed to us in experimentation is the maximal NR value produced in pure nitrogen. The values along the declining curve of NR towards its zero value—$NR_{E.P.}$—at total extinction can be at once obtained by difference when an OR curve has been adopted.

When we think of this NR curve in terms of its physiological determination, we realize how different is its quality from that of its fellow-component OR. The NR-max. value stands out from all the declining values as having alone a primary significance. It is a direct measure of an independent function—our NR function—which is in superficial aspect quite independent of oxygen. Moreover, this NR-max. is the direct determinant of the high initial TR value, being in fact identical with it.

The declining series of NR values along the NR curve has no primary significance. It merely marks the stages of gradual extinction of the fermentation function by the rise of oxygen supply in association with the aerobic respiratory function. These NR values are then unextinguished residues of the total NR function and will accordingly be labelled 'NR residues'. It is not the NR function which is the physiological determinant of their magnitude, but their magnitudes are an outcome of the extinguishing activity of the OR function.

Since the NR-residue values decline to zero at the E.P., they do not make a very prolonged contribution to the whole range of magnitudes of the TR curve. But for the low [O_2] region which is our present subject of study they are of course of great importance.

The locus of the extinction point. This conception has several attributes. Our first use of the term was just for marking the extinction of NR as a component of TR. We then passed to the view that it marks a definite break in the physiological control of respiration rate. In addition, we identified this *locus* with that of the upward inflexion at the minimum of the TR curve.

The physiological determinants of E.P. and the justification of the above attributes will be discussed in the next paper containing the later stages of our analysis. Here we only have need of E.P. as the locus of the zero value of NR.

(b) *Oxidative substitution as an index in resolution of* TR *values:* NR *deficits and* S *values*

The resolution of TR dealt with in the previous section can be formalized by the adoption of our first equation:

$$TR = OR + NR\text{-res.} \tag{1}$$

To formalize the significance of the relations between declining NR residues and increasing OR values we shall regard these phenomena as the expression of a function of correlated substitution in which as the oxygen supply increases, OR, measured as so many carbon units converted to CO_2 by oxidation, takes the place of NR, considered as so many units of carbon set free as CO_2 by fermentation. This function, thus viewed from the respiration side, we shall entitle *oxidative substitution*.

To express this functional activity we have to relate OR magnitudes to a new type of NR magnitude, namely, the NR that has been extinguished and is now recorded as in deficit. For these values we shall write NR-*deficit*. Such NR-def. values are measured as deficits from the maximal value of NR that is observable in complete absence of oxygen, where NR-def. must be zero.

These three types of NR values that we have to distinguish can be simply related by expression (2):

$$\text{NR-max.} = \text{NR-res.} + \text{NR-def.} \tag{2}$$

It follows that to obtain values of NR-def., which are the centre of theoretical interest in this substitution, it is necessary to make two separate observations on

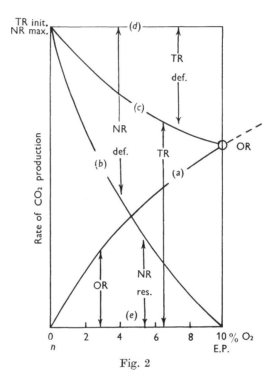

Fig. 2

a tissue, one in pure nitrogen to give NR-max., as well as one in the particular low $[O_2]$ under consideration, from the TR value in which we hope by analysis to extract the derivative NR-res. value.

We note in connexion with these three categories of NR values that we can recognize, graphically, similar categories for TR values and write:

$$\text{TR-initial} = \text{TR-res.} + \text{TR-def.} \tag{3}$$

This terminology will be found useful for calculations, though TR-initial is only a synonym for NR-max.

The relations of these magnitudes may usefully be presented in a generalized graphic form (fig. 2).

The ordinates are throughout rates of CO_2 production. The oxygen abscissa axis, (e), extends from zero $[O_2]$ to the $[O_2]$ that locates the extinction point. The OR component of TR, (a), rises throughout this region and continues on beyond the E.P., where it had reached the value of $OR_{E.P.}$ and henceforward constitutes the whole of the observed TR curve.

The NR-res. curve (b) starts over N at its maximal value and steadily declines to zero at E.P.

The TR curve, (c), starts at TR-initial (identical with NR-max.) and pursues its down course as a mixture of OR and NR till it reaches the inflexion point value, TR-min., at E.P.

Extending the TR-initial value as the horizontal line (d) we are able to plot deficit values downwards from it. The ordinate values between lines (d) and (c) give TR-def. values, while NR-def. values are indicated between (d) and (b).

The five values included in expressions (1), (2) and (3) are thus presented in fig. 2 as ordinates related graphically. It will be noted for future convenience when we develop the substitution relations of deficits that while NR-def. cannot be arrived at from the direct primary observations yet the TR-def. values are available, being merely the extent to which the successive TR values actually observed in low oxygen fall below the higher TR-initial value in nitrogen.

Considering fig. 2 geometrically we notice that one area has not been brought to account, namely, the cusp-shaped area between lines (a) and (c), which ends in a zero value at E.P. It is easily made out that this area repeats the ordinate values between (b) and (e) and therefore gives the measure of NR-res. all along the O_2 axis. To get from fig. 2 the most direct simple picture of the quantitative relations without any overlap of areas, we should remove line (b) altogether. Then we have left three areas whose individual vertical heights represent respectively OR, NR-res. and TR-def., while their sum has always the constant value of TR-initial representing the whole rectangle.

We now seek to extend our sequence of expressions in order to link up the relation of the observable TR-def. to the theoretically desirable NR-def. This is carried through as follows:

Since NR-max. of (2) is identical with TR-initial of (3) we can write

$$\text{TR-def.} + \text{TR-res.} = \text{NR-def.} + \text{NR-res.} \tag{4}$$

Substituting for NR-res. in (4) by expression (1) we arrive at

$$\text{TR-def.} = \text{NR-def.} - \text{OR.} \tag{5}$$

We note then that for calculating the relations shown in fig. 2 all along the curves, NR-def. is always larger than TR-def. by the magnitude of OR in the particular low O_2 under consideration.

The last stage in the development of these quantity relations for the numerical evaluation of the oxidative substitution relation takes the form of adopting an 'oxidative substitution coefficient' S for the ratio of the correlated values involved in substitution.

For this purpose we may write

$$\frac{\text{NR-def.}}{\text{OR}} = S, \tag{6}$$

or

$$\text{NR-def.} = \text{OR} \times S. \tag{6a}$$

We can thus check out the substitution relations of any proposed resolution of TR curves into OR + NR curves and see to what extent the suggested values give a fluctuating or constant ratio for S.

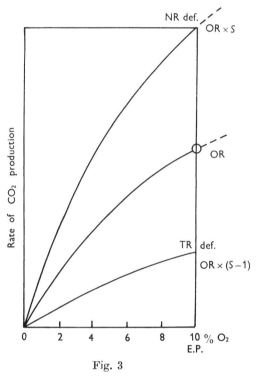

Fig. 3

As an alternative expression making use of the directly obtainable TR-def. value we can employ

$$\frac{\text{TR-def.}}{\text{OR}} = S - 1; \tag{7}$$

or

$$\text{TR-def.} = \text{OR} \times (S - 1). \tag{7a}$$

In fig. 3 we present the three curves OR, TR-def. and NR-def., showing how they all increase, in similitude, with movement along the $[O_2]$ axis, a similitude which is complete only if the reckoned S values should come out identical at all stages of the curves. This extraction and contemplation of S values along the curve system provides a new test for the rationality of OR and NR component curves, from nitrogen up to the E.P. line.

Fig. 3, as presented here, with NR-def. above OR and TR-def. below OR, implies that S has a greater value than 1·00 (here S is presented as equal 1·66), so that OR × S lies above OR but its value is also by implication less than 2·00 because $(S-1)$ OR is brought down to lie below the OR curve. Actual values of S will come up for survey in § 2.

The evaluation of the coefficient S as just depicted is here dependent upon accurate resolution of observed TR curves into their OR and NR components. We had to search for possible opportunities of getting more objective values of S than the only ones that, at first, appeared to be attainable, namely, S values arrived at by subjective graphic resolution of TR.

In time, it occurred to us that if we were assured of our location of E.P. on the $[O_2]$ axis, under the TR curves of fig. 1, then we could arrive directly at some value of S. This follows because at E.P. by definition the whole of NR is just extinguished and therefore the value of NR-def. is given numerically by the observed value of NR-max. in nitrogen; while, equally by definition, the minimal TR value which characterizes the E.P. is just the OR value that is substituting this deficit of NR-max. We can therefore write

$$\frac{\text{TR-initial (as observed in nitrogen)}}{\text{TR-min. (as observed at E.P.)}} = \frac{\text{NR-def. (at E.P.)}}{\text{OR (at E.P.)}} = S. \qquad (8)$$

In the next section such *observed* values of S will be reported and compared with sets of S values derived from the resolution of the four TR curves that constituted the first objective of our analytic attack.

We have now sketched the quantitative relations of OR and NR as components of TR, and indicated the value of S as a rationalizer of OR/NR relations. As an additional aid to analysis we have also introduced the conception of a definite extinction point on the $[O_2]$ axis, where there is a break in the physiology of TR by the extinction of the NR-res.

Before applying these methods to our actual respiration data we must provide an outline of the interference that may result from the interaction of what we have called the *carbon function*.

(c) *Elimination of 'carbon grade' distortions by standardization:* *he ontogenetic family of* TR *curves*

In the previous pages our attention was concentrated chiefly on the striking form of any one single TR curve. In the present section we bring up for consideration the relations between the four constituents of what we call our ontogenetic family of TR curves. They represent a drifting series of stages through which each apple progresses as it develops towards senescence in cool storage. Some functional determinant must be altering with senescence to account for the regular change of form of TR curves.

We must first stress the fact that this set of curves does not record absolute values of CO_2 production, but only relative rates. We are contemplating a *standardized* set in which the TR value in air is always taken as a unitary standard of reference to which all other values are relative. Thus all the values spoken of as TR values throughout this survey are really TR/OR values and therefore their resolved components OR and NR are numerically OR/OR_A and NR/OR_A values.

As this standardization for the whole seasonal drift conceals from us evidence about the magnitude of senescent drift of OR_A itself, it is clearly a handicap retarding full analysis. We have therefore now to point out why standardization of some sort is essential for these apple data, to avoid still greater evils.

When attempting to carry out, on apple tissue, an extensive study of oxidation effects (such as the oxidative substitution of OR for NR through a range of low $[O_2]$) the investigator meets a fundamental source of trouble as soon as he undertakes experiments of long duration. This trouble we may define as distortions of value due to *changes of carbon grade*, that occur spontaneously in the metabolism of the tissues. Apples in storage undergo a progressive lowering of the availability of their internal supplies of carbon metabolites presenting a phenomenon that we call, loosely, progressive starvation.

The respiration data for the present work could only be collected slowly and the experiments extend over months, during which the general pitch of carbon metabolism first rose and then fell. In order to draw comparisons between oxidative behaviour at remote dates it is necessary to introduce some form of *carbon standard*.

We may give an illustration of this carbon trouble. Suppose we are measuring respiration in low oxygen in which both NR and OR functions are proceeding simultaneously, and afterwards a second repetition of the same experiment, carried out at a much later date, gives definitely less OR than the earlier one. This might be due to two different causes, either a change in oxidative activity or to a spontaneous starvation fall in carbon supply. We could recognize which cause was at work because with lowered carbon grade the NR production would in fact fall as well as the OR, while with unchanged carbon grade but fall of oxidative substitution efficiency the NR would rise when the OR fell. Recognizing the trouble of this carbon-grade distortion, we have the problem of devising a schematic way of treating it, so as to segregate its activity. Our formal attempt consists in collecting the oxidative factors inside a bracket and putting the carbon-grade factor outside so as to maintain a sharp distinction. In its simplest form we may write this as

$$\text{carbon grade} \left[\begin{array}{c} \text{oxidative substitution} \\ (NR)\, v.\, (OR) \end{array} \right]$$

a form which implies that change of carbon grade affects similarly both NR and OR inside the bracket, so that change of it alters the totality of TR.

As our objective at present is not the detailed study of carbon grades but the analysis of oxidative functions we push to one side the carbon situation by adopting some uniform standard of carbon grade to which all studies of variation between the components inside the bracket may be referred.

The procedure in practice would be that each observation of the effects of low oxygen in altering NR $v.$ OR would be associated with a second observation, simultaneous in ideal cases on an identical tissue in the selected oxygen concentration adopted to give the carbon standard as unit of reference. The low oxygen result would be expressed as a ratio to the standard oxygen result, as unit.

We have next to consider what oxygen condition should be chosen to provide the carbon standard. Obviously the condition should lie outside the range of so-called 'low oxygen' which shifts the components inversely. It should be an oxygen concentration which is, at least, so extreme that substitution is complete—one way or the other—the total CO_2 being either all NR or all OR. Nitrogen, i.e. zero oxygen, is the only condition to secure the state that TR is all NR, while a good range of high oxygen concentrations will give the state of all OR.

For all *normal* states of apple tissue, 21 % O_2 gives this 'all OR' state, since substitution is complete with about half this concentration. The intense convenience of adopting air as a standard condition has also enforced its employment as a universal standard. It must be noted, however, that it is only acceptable when the E.P. locus, where the NR component ceases, is known to be definitely below air.

We shall adopt the air standard as our primary one, but shall occasionally present data on the nitrogen standard. A glance forward at pp. 154–5 will reveal how very different is the appearance of the family relationship of our four onto-genetic curves, according as the TR curves are set out on an air standard (fig. 5) or a nitrogen standard (fig. 6).

2. COMPUTATION OF OR, NR AND S VALUES FOR THE RESPIRATION OF APPLES IN LOW OXYGEN

(d) *First approximate resolution of* TR *by graphic interpolation*

We have now to proceed to the actual resolution of the ontogenetic family of TR curves into a family of OR curves and a family of NR curves. As the OR curves alone represent a primary function, our actual task is to construct the family of OR curves, after which NR can be derived by difference.

As was stated in the general discussion in the previous section the TR curves are a mixture of OR + NR up to their inflexion point which locates E.P. The further range of the TR curve beyond the inflexion consists wholly of OR so that a considerable, high region of the OR curves is revealed to us. The curves in fig. 1 lie close together in this region because they are standardized OR/OR_A curves on which the standard of unity at 21 % O_2 is imposed. We note that the grade of curvature within the set of revealed OR curves differs in the region of the minimum showing less curvature at the more senescent stage of ontogeny, and we pass on to our special problem of arriving at a satisfactory course for the masked OR curves below the E.P.

For making the graphic interpolations which we shall put forward in this section we prepared a separate large-scale graph of each of our four observed TR curves. These graphs shown on a reduced scale in fig. 4a and b form the starting point for the present section. In them will be found the various OR and NR curves that are to be discussed as possible components. We decided that the first approximation would be to try out two widely separated tracks for the interpolated OR curve. These were intended to be such extreme forms that the true OR track could not be expected to lie outside them in its course from the origin to the TR-min.

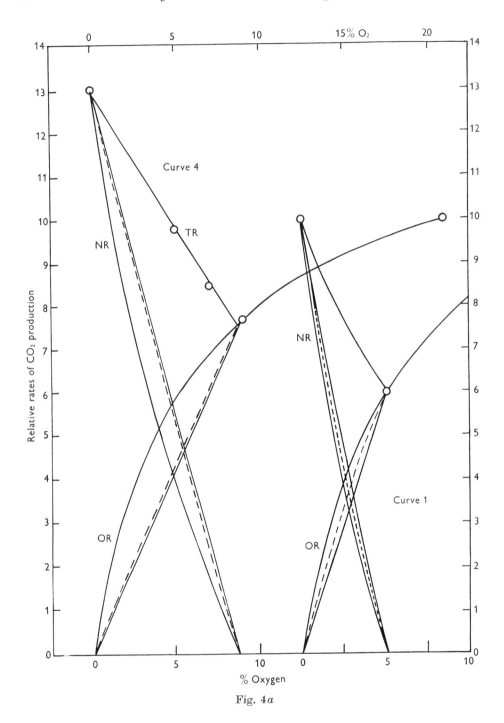

Fig. 4a

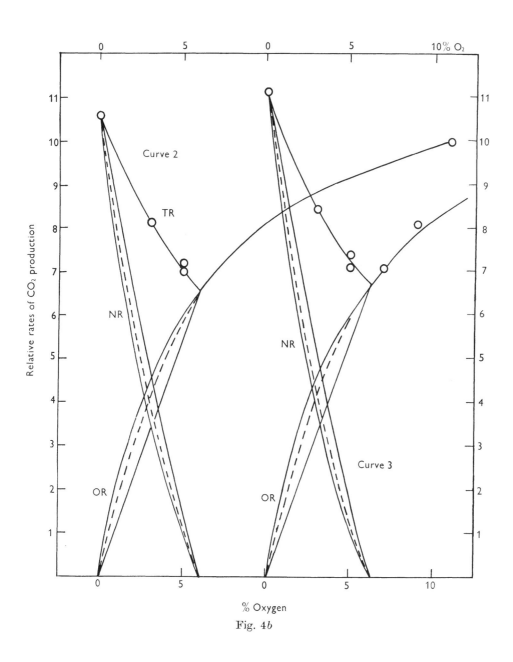

Fig. 4b

Guided by the considerations of OR form set out in § 1 we prepared as one possible extreme case a perfectly rectilinear course all the way from the origin to the E.P. Such lines were drawn on each of the four graphs, while for the other extreme there were drawn, freehand, curvilinear tracks carrying back the revealed part of OR

Table 1. *Analysis of the family of four* TR *curves*

TR curves		OR with curvilinear form				OR with rectilinear form			
1 O_2 %	2 TR	3 OR	4 NR	5 NR-def.	6 S	7 OR	8 NR	9 NR-def.	10 S
TR curve 1									
0	10·0	0·0	10·0	—	—	0·0	10·0	—	—
1	9·0	2·0	7·0	3·0	1·50	1·2	7·8	2·2	1·83
2	8·1	3·4	4·7	5·3	1·56	2·4	5·7	4·3	1·80
3	7·3	4·45	2·85	7·15	1·61	3·6	3·7	6·3	1·75
4	6·6	5·35	1·25	8·75	1·63	4·8	1·8	8·2	1·71
5	6·0	6·0	0·0	10·0	**1·66**	6·0	0·0	10·0	**1·66**
TR curve 2									
0	10·6	0·0	10·6	—	—	0·0	10·6	—	—
1	9·65	2·0	7·65	2·95	1·47	1·1	8·55	2·05	1·86
2	8·8	3·4	5·4	5·2	1·53	2·2	6·6	4·0	1·82
3	8·1	4·45	3·65	6·95	1·56	3·3	4·8	5·8	1·76
4	7·5	5·35	2·15	8·45	1·58	4·35	3·15	7·45	1·71
5	7·0	6·0	1·0	9·6	1·60	5·45	1·55	9·05	1·66
6	6·55	6·55	0·0	10·6	**1·62**	6·55	0·0	10·6	**1·62**
TR curve 3									
0	11·1	0·0	11·1	—	—	0·0	11·1	—	—
1	10·1	2·0	8·1	3·0	1·50	1·07	9·04	2·06	1·92
2	9·25	3·40	5·85	5·25	1·55	2·14	7·11	4·0	1·87
3	8·45	4·50	3·95	7·15	1·59	3·21	5·24	5·86	1·82
4	7·85	5·45	2·30	8·8	1·60	4·28	3·57	7·53	1·76
5	7·25	6·10	1·15	9·95	1·63	5·35	1·90	9·2	1·71
6·3	6·75	6·75	0·0	11·1	**1·64**	6·75	0·0	11·1	**1·64**
TR curve 4									
0	13·0	0·0	13·0	—	—	0·0	13·0	—	—
1	12·35	2·0	10·35	2·65	1·32	0·85	11·5	1·50	1·76
2	11·75	3·4	8·35	4·65	1·37	1·70	10·05	2·95	1·73
3	11·05	4·4	6·65	6·35	1·44	2·6	8·45	4·55	1·75
4	10·45	5·25	5·2	7·8	1·49	3·45	7·0	6·0	1·74
5	9·8	5·9	3·9	9·1	1·54	4·3	5·5	7·5	1·74
6	9·2	6·4	2·8	10·2	1·60	5·2	4·0	9·0	1·73
7	8·6	6·9	1·7	11·3	1·64	6·05	2·55	11·45	1·73
8	7·95	7·35	0·6	12·4	1·69	6·9	1·05	11·95	1·73
8·7	7·55	7·55	0·0	13·0	**1·72**	7·55	0·0	13·0	**1·72**

to give the greatest acceptable curvature near the origin. Here then we have two graphic interpolations as possible OR courses for each TR curve; a third course, which will be brought up later, will be seen to have been added to each figure, lying, as a broken track, between the other two.

Having drawn this variety of OR curves freehand it is essential that we should take out their numerical values as the basis for computing the associated NR-res. curves and other derivative values. Table 1 presents us with the first group of values.

In this table the four TR curves 1–4 are dealt with in succession, the values arising from the adoption of the graphic curvilinear form of OR being on the left half of the table, those for the rectilinear OR being on the right. A series of values is given for each curve from 0 % O_2 advancing by steps of 1 % on to the extinction point as marked by the zero values of NR in columns 4 and 8. These E.P. values have been already, by graphic study of the position of the minims on the TR curves, located at the following oxygen percentages: curve 1 at 5·0 %; 2 at 6·0 %; 3 at 6·3 % and 4 at 8·7 % (see Paper V, p. 121).

The derivation of the separate columns of table 1 should be obvious. The external O_2 concentration occurs in column 1 while the observed TR values are set out in column 2. Then in association with the two forms of proposed OR curves entered in columns 3 and 7 follow the two related sets of NR-res. values in columns 4 and 8. The two NR-res. values in zero oxygen are identical within each pair of cases as there is no OR, and NR is here 'maximal'. The rest of the NR values in these columns 4 and 8 are NR-res. still left unsubstituted. In the graphs of fig. 4 these NR curves all decline rapidly from the NR-max. value in nitrogen down to zero value at the corresponding E.P. which is located at the last value of O_2 percentage entered in column 1 for each section of the table.

(e) *Second approximation: computation of pairs of* OR + NR *values correlated through S values*

We may now proceed to examine these various colligate pairs of OR + NR by the test of computing their substitutionary relations, as shown by the value of the oxidative substitution coefficient that each pair provides.

Columns 5 and 9 contain the NR-def. values, measured by NR-max. less column 4, and giving the values of the NR that has disappeared owing to substitution by OR. Relating these to the corresponding OR values by use of expression (6), $\dfrac{\text{NR-def.}}{\text{OR}} = S$, we obtain the series of individual S values that are set out in columns 6 and 10.

On examining the values of S in these columns it will be seen that they range from 1·32 to 1·92 over the whole table, but that the scatter is much smaller within each single subtable. The S values for rectilinear OR are always higher than for curvilinear, except at the last values in each subtable at the level of the E.P. concentration of oxygen.

Here the values are identical, and these values (printed in heavy type) have a different status from the rest, because in their case the S ratio is based upon values which occur in the primary TR curves of fig. 1 and not upon interpolated values of component curves.

The four values we shall call *direct S* values. Turning back to p. 145 we find expression (8) given as a method of calculating one S value directly from each TR curve. On account of their importance the four values may here be set out in the text:

$$\text{Curve 1, } \frac{10\cdot0}{6\cdot0} = 1\cdot66; \quad \text{Curve 2, } \frac{10\cdot6}{6\cdot55} = 1\cdot62;$$

$$\text{Curve 3, } \frac{11\cdot1}{6\cdot75} = 1\cdot64; \quad \text{Curve 4, } \frac{13\cdot0}{7\cdot55} = 1\cdot72.$$

There is thus revealed a very close agreement between the four direct S values; practical identity between the three anterior to the most senescent stage. This small variation certainly suggests that the phenomenon of oxidative substitution proceeds on very uniform lines in spite of marked general ontogenetic drift.

In contrast with the uniformity of direct S values we notice a marked drift inside the blocks of 'interpolated' S values, so that the values at the level of 1% O_2 are often far removed from coincidence with the direct values that close the series.

This array of values will enable us to survey the extent to which the S value is affected by the choice of graphic form for the OR component. One regularity that comes out clearly is that for any 'rectilinear' OR set the values of S always decrease as we pass towards the E.P., while for the 'curvilinear' OR sets they always increase. Clearly then it should be possible to pick a form of OR curve intermediate between the two selected extreme forms which would give identical S values throughout its own series. It will be noticed further that in one of the eight curves, but only one, we have already a practically constant S value. Curve IV with rectilinear OR shows no drift of decrease; for all the S values fall between $1\cdot76$ and $1\cdot72$.

We may provisionally assume that constancy of S values all through a substitution series between OR and NR is really an index in favour of the correctness of the interpolation. If so, it becomes of great interest to examine how such interpolated curves would compare (in fig. 4) with the extreme forms already entered there.

We may therefore proceed to calculate such curves, using as the only acceptable S values for the four cases the recorded direct values $1\cdot66$, $1\cdot62$, $1\cdot64$ and $1\cdot72$. The calculation is on p. 144:

$$\text{OR} = \text{TR-def.}/(S-1). \tag{7a}$$

The OR values set out in column 4 of table 2 are derived in this way from the TR-def. values of column 3. In column 5 we set out the NR-res. values derived by subtracting the calculated OR from TR values of column 2. The new pairs of OR and NR thus calculated are inserted in fig. 4 as the middle curves (in broken lines) of each triple group. Nr-def. values could be easily derived from column 5 of the table, but they are not included in this printed table.

This manipulation of curves shows that value attaches to the oxidative substitution coefficient S as a weapon of analysis should it be established that it is really a constant, at all events for a given physiological state. The significance of its constancy through such a long set of data as is provided by curve IV extending from 1 to $8\cdot7\%$ O_2 would lie in the deduction that any partial substitution of NR by OR is independent of the fraction of the amount of NR that has been already substituted.

We may now say that by exploitation of S we have provided ourselves with a set of OR and NR curves which conform to a numerical substitution standard and are not merely freehand products. They constitute a correlated set instead of a graphic set. In curves 1, 2, 3 of fig. 4 the new values lie neatly between the curves we examined as extreme forms, namely, the curvilinear and the rectilinear form; while for curve 4 the new OR curve is hardly distinguishable from the old rectilinear form.

Table 2. *Calculation of* OR *curves on basis of constant S values*

TR curve 1 adopting $S = 1 \cdot 66$					TR curve 3 adopting $S = 1 \cdot 64$				
1 O_2 %	2 TR	3 TR-def.	4 OR-calc.	5 NR-res.	1 O_2 %	2 TR	3 TR-def.	4 OR-calc.	5 NR-res.
0	10·0	0·0	0·0	10·0	0	11·1	0·0	0·0	11·1
1	9·0	1·0	1·5	7·5	1	10·1	1·0	1·56	8·54
2	8·1	1·9	2·85	5·25	2	9·25	1·85	2·89	6·36
3	7·3	2·7	4·05	3·25	3	8·45	2·65	4·14	4·31
4	6·6	3·4	5·1	1·5	4	7·85	3·25	5·08	2·78
5	6·0	4·0	6·0	0·0	5	7·25	3·85	5·98	1·27
					6·3	6·75	4·35	6·75	0·0

TR curve 2 adopting $S = 1 \cdot 62$					TR curve 4 adopting $S = 1 \cdot 72$				
0	10·6	0·0	0·0	10·6	0	13·0	0·0	0·0	13·0
1	9·65	0·95	1·53	8·1	1	12·35	0·65	0·9	11·45
2	8·8	1·8	2·9	5·9	2	11·75	1·25	1·73	10·02
3	8·1	2·5	4·03	4·1	3	11·05	1·95	2·71	8·34
4	7·5	3·1	5·0	2·5	4	10·45	2·55	3·54	6·91
5	7·0	3·6	5·8	1·2	5	9·8	3·2	4·44	5·36
6	6·55	4·05	6·55	0·0	6	9·2	3·8	5·28	3·92
					7	8·6	4·4	6·11	2·49
					8	7·95	5·05	7·01	0·94
					8·7	7·55	5·45	7·55	0·0

As now derived the set of OR curves shows a drift of form from curvature in I, II and III to a straight line in IV. These we may adopt as the best attested set that we can bring forward when we work through the conception that, for a given physiological state, S is constant throughout the successive steps of substitution from beginning to end. This set consists of three curved and one straight line, so there is a definite drift of form in this region with the progress of ontogeny and the drift is a downward one so that OR curve IV lies below the curved earlier forms and starts at a smaller angle from the origin.

(f) *Survey of resolved* OR *curves on air and nitrogen standards*

Having pushed our analysis of the components of the TR curves as far as can be managed on the present lines, we may now discard the first approximations and adopt the 'correlated' components as the expression of the OR function. For the presentation of our survey of the situation in graphic form we have to make preliminary decisions about the carbon standard to be adopted.

On p. 147 we forecast the importance of surveying mixed values from the nitrogen as well as the air standard. This survey is now carried out in figs. 5 and 6. Let us first glance at fig. 5 for the air standard. Here the TR curves reappear exactly as in fig. 1, but now associated with their respective OR curves. Of these, the tracks of curves I, II and III are practically indistinguishable and have a curved

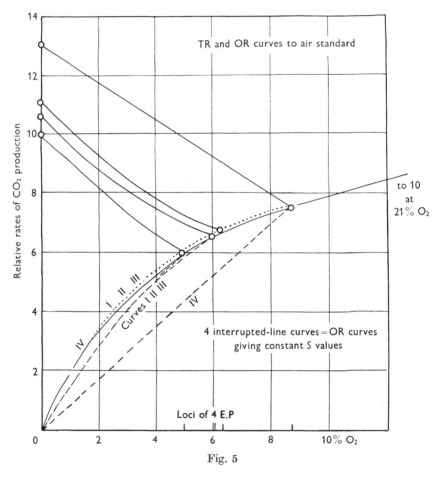

Fig. 5

form, while curve IV runs well apart in a straight line from the origin to its $OR_{E.P.}$ value at its TR-min. On this standard the picture given is that the outstanding effect of ontogenetic drift is a big increase in the values of NR-max.

The appearance presented by fig. 6 on the nitrogen standard is, however, completely different. Here we find NR-max. (TR-initial) presented, by definition, as unity in all four cases, while the four OR curves diverge steadily from the origin till in air they are as far apart as the range from 10·0 to 7·7. The least senescent OR curve is on top and the most senescent is lowest. Taking these two figures together and asking which gives the true picture of the effect of ontogenetic drift, the answer is of course that speaking by the information that is set out in the present paper we cannot yet tell. The effect of the drift may be wholly to raise NR-max. and leave OR unaltered, or it may be to leave NR-max. unaltered and solely to depress OR. It requires special studies of absolute values to decide between these extremes. When these are available we shall be able to decide whether OR is so much depressed or whether the truth is not something between the two extremes so that NR max. rises a little and OR falls a little.

This present uncertainty arises, of course, as the drawback of using standardized data. No simple standard can be quite satisfactory for a curve of the composite nature of these TR curves. Either standard masks a scatter at its own end of the axis and exaggerates scatter at the other end.

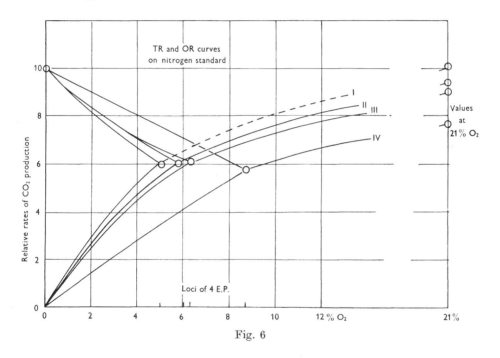

Fig. 6

Our temporary position on this matter has to be that we can only think in terms of a *relative determinant*. This determinant should be defined as 'the differential drift apart between OR and NR', so as to provide indifferently for upward drift of NR or downward drift of OR alone, or for the case that both rise or both fall though to different degrees. The relative drift having been now determined, then a measure of the absolute drift of NR or of OR will allow us to distribute the whole effect. Such data have been collected but not yet adequately organized for publication.

While discussing these two standards it may be pointed out that, with either carbon standard, the curves are extended over an oxygen axis which is the concentration of *external* oxygen. It follows that the determinants of drift along the axis are not purely metabolic functions; but the biological function of diffusion resistance is interwoven with them. Both increase of diffusion resistance and decrease of biochemical oxidative efficiency can lower the slope of the OR line.

At present we lack the data to discriminate between these two subdeterminants. A minor point that may now be mentioned is the relation of curvature of form to standardization.

Determinant of curvilinearity of OR. In contrast with distortion of pitch and levels, there is one determinant of absolute curve forms which is brought out more clearly in standardized curves, and this is the grade of curvilinearity as distinguished from rectilinearity shown by the set of OR curves derived from the four TR curves. When the various OR curves, in addition to coinciding, necessarily, at their zero origins, are made to coincide in air by standardization, then any differences of curvature between become outstanding and we see in the air standards (fig. 5) that the early OR curves are curvilinear while the most senescent one is practically straight. Though the greater curvature of the less senescent OR form is thus clearly revealed we cannot tell merely from inspection of the air-standard graph whether this younger curved OR had an absolutely higher pitch at its air end than did the older straight OR, or vice versa.

We may bring this second section of our study to an end by a formal statement of the range of types of relation between OR and NR that we have encountered. We can distinguish as many as three definable relations, which are of the form of (1) sympathy, (2) drifting apart, and (3) antipathy.

The chief talk in this paper has been about (3) the inverse correlation of OR *v.* NR in oxidative substitution. This seems to be a strict inverse correlation measured by the coefficient S. It is this relation which dominates the CO_2 phenomena in the region of low oxygen between N and E.P.

Differing in nature from this is relation (1) which is correlated with 'carbon grade', where the general phenomenon is that as carbon grade rises or falls, as in starvation and general nutrition, so both NR and OR rise or fall together. This concerns the general broad aspects of high or low respiration rates.

Superposed on this major relation of sympathy we have as (2) the minor drift apart of OR and NR that is associated with ontogenetic drift.

In the Introduction to this paper we spoke of five physiological determinants that could be identified at work and gave them the symbols OR, NR, S, E.P. and C. We have now made most of the points that have arisen about OR and NR and there has been some exposition of the carbon grade (C). The further exposition will develop the carbon grade aspects of respiration and carry analysis more deeply into the phenomenon of E.P. and S.

VIII

THE PHYSIOLOGICAL DETERMINANTS OF CARBON DIOXIDE PRODUCTION BY APPLES IN LOW CONCENTRATIONS OF OXYGEN*

CONTENTS

1. THE MAGNITUDE OF THE COEFFICIENT OF OXIDATIVE SUBSTITUTION —S—, AND ITS DETERMINANT EFFECTS IN RESPIRATION

Introduction. In the previous paper of this series (Paper VII) there were carried out the opening stages of our analysis of the empirical data of the respiration of apples in low $[O_2]$. This analysis led us at an early stage to bring forward the view that behind the graphic features there must be at work a special cause controlling the relations in the intermediate metabolism preparing the products of respiration and of fermentation. For this we proposed the name of oxidative substitution, and associated with it a special coefficient, S. We may recapitulate, in a paragraph, the exact significance that is to be attached to this coefficient.

It was recognized that the activity of the respiratory function, R, can prevent the appearance in tissues of the products of the (alcoholic) fermentation function, F. And, further, that the amount of R product put 'in deficit' in this way has a definite quantitative relation to the number of units of product of R, actively substituting this F. When our survey of the activity of F and of R is only made in terms of numbers of CO_2 units involved, we find, empirically, that for one unit of OR arising, a greater number of units of NR will be extinguished. In certain well-investigated cases we found that 1·66 units of NR are substituted by one unit of

---

* This paper has not been published previously. No authors were named in the typescript, which was dated 18 Aug. 1935.

OR. To express this relation we gave our substitution coefficient the form of $\dfrac{\text{units of NR-def.}}{\text{units of OR}}$ substituting, and hence S has the magnitude here of 1·66. Other variant values (all above unity) have been encountered in parts of our extended survey of the respiration of apples.

Similar high values of S have been found in certain other respiring plant tissues, but on the contrary some types of plants show values below unity. These differences are presumably linked up with the nature of the *total metabolism* involved in F and in R during substitution. For the present survey we must limit our exposition strictly to the phenomena and relations of the volatile CO_2 production.

At the shallow depth to which we are now pushing our analysis S can claim the attributes of a primary function and will be treated as a biological determinant and thus an independent variable.

To the consequences that attend variations of magnitude of S we must pay profound attention, for this may alter the whole facies of respiration data in low concentrations of oxygen. The special initial form of the TR curve in apples will be held to be directly due to the high values of S in this tissue; with low values of S this striking characteristic would be absent.

This relation of TR slope to S values forms the second topic for discussion in the present section. Before taking that up we must expand and generalize the relations between NR and OR magnitudes that are controlled by S.

(a) *The full interrelation between* NR *and* OR *values*

For fig. 1 we have prepared a diagram which presents in its most general form the control of substitutional relations by the coefficient S. The figure shows two rectangles of different heights but superposed as regards their common base-line, AB. Each rectangle is fitted with a pair of curves. The smaller rectangle $ABEF$ and its related curves are drawn in continuous lines, and this system stands for the respiration function of apple tissues; while the taller rectangle, $ABCD$, drawn, with its pair of curves, in dotted lines stands for the fermentation function of the same tissue. The vertical sides are CO_2 ordinates of functional rate, NR and OR, while AB is the abscissa axis of $[O_2]$, extending from zero O_2 at A to the extinction point value E.P. at B.

The horizontal lines EF and CD represent the upper limiting ordinate value of each function, so that CD measures the initial TR value defined as NR-max. while EF gives the pitch of the OR value that completes the substitution of NR-max. This value, which has been known as TR-min., will in future be defined as $OR_{\text{E.P.}}$. It is a sort of maximum OR for the substitutional system, though outside this closed system OR goes on rising (from F to G in the diagram) till it presently reaches OR_A.

We may now turn to the four enclosed curves. From what we have just said the curve AF is evidently the rising curve of OR, while CB is the falling curve of NR-res. reaching zero at B. The other two curves are drawn geometrically as similar in trend and proportional throughout to their fellows sharing an origin with them. Functionally AD will be the curve of rising NR-def., while EB will be the curve of a declining attribute of OR which might be called an $OR_{E.P.}$ 'deficit', being zero at B when the full $OR_{E.P.}$ is actively substituting. In the diagram the height AC is 1·5 times the height AE which decides the value of S for the whole binary system, since $S = \dfrac{\text{NR-max}}{OR_{E.P.}}$. Geometrically this S value of 1·5 must hold throughout for the relation of corresponding points on each pair of curves, $\dfrac{\text{curve } CB}{\text{curve } EB}$ and $\dfrac{\text{curve } AD}{\text{curve } AF}$.

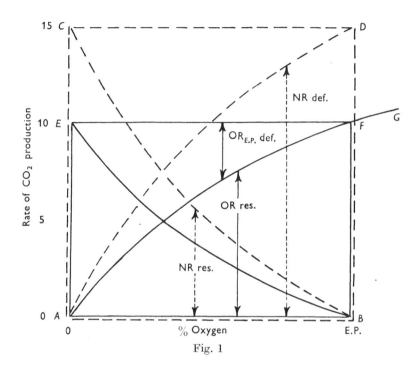

Fig. 1

In addition, certain geometric identities are inherent in each of the two rectangles. In the larger fermentation function the series of ordinate values between the line AB and the curve CB are identical with those between line CD and curve AD, both series representing the magnitudes of NR-res.; while in the same rectangle the values between AB and AD are identical with those between CD and CB, both representing NR-def.

Similarly for the smaller rectangle of the R function, there are two like pairs of identities, (1) EB over AB with AF below EF called the OR-def., and (2) AF over AB with EB below EF representing the substituting OR (labelled, for geometric simplicity, OR-res., while functionally it should be OR-substituting).

From the geometrical similarities indicated by this diagram we can set out three expressions of NR and OR relations governed by S:

    (*a*) NR-max. $= OR_{E.P.} \times S$,
    (*b*) NR-res. $= $ OR-def. $\times S$,
    (*c*) NR-def. $= $ OR-res. $\times S$.

To these we may add the two expressions (*d*) and (*e*) for the summations of various NR and OR magnitudes, which are applicable solely within the OR limits of this schema which maximates at $OR_{E.P.}$:

(*d*) NR-max. = NR-res. + NR-def.,

(*e*) $OR_{E.P.}$ = OR-res. + OR-def.

Here then we have the whole picture of substitution, as it unrolls in apples where the NR dimensions are all of greater magnitude than the substituting OR dimensions. It is interesting to note now that this diagram can equally be applied to the alien organization of respiration that holds in plants in which the NR dimensions are smaller than the substituting OR dimensions. For such application we have only to decide that the tall rectangle shall stand for the *R* function and the short one for the *F* function. The geometry will be as before, though all the functional labels will be shifted over. Therefore, if *S* retains its definition (with NR in numerator and OR in denominator) then throughout the new schema we shall have $S = 1\cdot0/1\cdot5 = 0\cdot66$. The diagram will thus serve also to illustrate the case of a tissue with the *S* value well below unity.

(*b*) *The magnitude of S as a determinant of* TR *slope*

The major complication of all respiration curves as set out over an abscissa of $[O_2]$ is that in the initial region neither the course of the NR curve nor of the OR curve is revealed, but only the course of their aggregate, the TR curve. The complexity of this can only be met by careful analysis, and it will be convenient to develop this analysis to a wider issue as a direct outcome of the study of *S* relations in the previous section. Expression (*f*) makes a formal introduction to the dual components of TR:

(*f*) TR = OR-res. + NR-res.

Since the magnitude of *S* determines the relative quantitative importance of the two components in the make-up of the sum we shall consider a pair of extreme cases together, in one of which *S* has the high value of $1\cdot66$, while in the other *S* has the hypothetical very low value of $0\cdot33$. It would be too confusing for fig. 1 to introduce the TR slopes needed for our present purpose, so we have drawn up fig. 2, in which all the 'curves' that we are concerned with are given strict rectilinearity for convenience of presentation. This figure sets out for each case three rectangles for the respective $OR_{E.P.}$, NR and TR functions, while the diagonals divide NR and $OR_{E.P.}$ into the complementary 'residue' and 'deficit' parts.

The $OR_{E.P.}$ rectangles occupy the upper part of the figure, being both given a common ordinate value of $3\cdot0$ CO_2 units. Here the rising diagonal represents the rising set of $OR_{E.P.}$-res. values from zero at the origin to $3\cdot0$ at the E.P. locus. The OR-def. values, for the part of $OR_{E.P.}$ which is not actively substituting, are above the diagonal, being maximal at the origin and zero at E.P. The steepness of slope is measured by the rise of $3\cdot0$ units over the extension of the $[O_2]$ axis of the diagram. Both OR slopes then are identical and independent of *S*, and may be defined as $+3\cdot0$ units.

Turning now to the NR rectangles, below them, we see that one is drawn very tall and the other very short. Since NR-max. is determined by $OR_{E.P.} \times S$ it follows that the value of NR-max. is 5·0 for the high S and 1·0 for the low S. From these initial values the NR-res. values must decline in a straight line to reach zero. These are the values below the diagonal while the increasing NR-def. values will lie above the diagonal.

The NR-res. slopes are in both cases declining towards the E.P. and, in contrast with the OR slopes, are therefore negative. The steepness is quite different for the two cases. For high S, the NR-res. slope is $-5\cdot0$ units, for low S it is $-1\cdot0$ units.

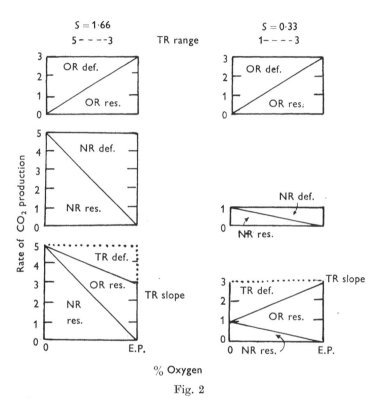

Fig. 2

We can now build up the TR rectangles and the TR slopes which must be the algebraic sum of the slopes of OR-res. and NR-res. Thus we get

With $S = 1\cdot66$: TR slope = [OR slope ($+3$) and NR slope (-5)] = -2.
With $S = 0\cdot33$: TR slope = [OR slope ($+3$) and NR slope (-1)] = $+2$.

The upper line presents the case as found in apples with marked downward (negative) slope of TR, from initial value 5·0 to final 3·0. For the lower case, with low S, we find the same steepness of slope 2 units but now a rising slope from initial value 1·0 to final 3·0.

The dotted line at the tops of the lowest rectangles shows that in the matter of TR-def. the slopes would be the reverse of the TR values, i.e. $+2$ and -2 respectively.

We have now demonstrated the way in which S values contribute to determining the slopes of all initial TR curves whether these curves are downward or upward in their slope. We shall make further contact with this effect when we survey the determinants of inflexion form in their totality of action.

(c) *Computation of S values without predetermination of* $OR_{E.P.}$ *and E.P. loci*

In this analysis of respiratory relations a great many matters turn upon the value of the coefficient S, and we are therefore anxious to bring forward the simplest experimental technique for arriving at this value. For the group of data which is analysed in the present paper we have no determinations of oxygen consumption but only of CO_2 output. We are therefore at a disadvantage and forced to adopt indirect empirical methods of approach, such as deciding the E.P. locus on a curve in order to identify that particular OR value, labelled $OR_{E.P.}$, which has the utility of indicating the magnitude of S through the expression

$$S = \frac{\text{NR-max.}}{OR_{E.P.}} \ .$$

It is therefore satisfactory to realize that S values can be arrived at much more directly when our data include a measurement of O_2 intake, or even a measurement of the respiratory quotient R.Q. giving the ratio $\dfrac{CO_2 \text{ output}}{O_2 \text{ intake}}$. Such measurements will be provided abundantly in the paper which follows this,* and it seems worth while in relation to the general theory of S relations to add herewith, as an Appendix, the simple procedure for computing S values on the minimum of directly observed gas-exchange data. For this purpose we need two experiments one which gives the CO_2 production in nitrogen, our adapted NR-max., and a second experiment in any low oxygen concentration which is low enough to ensure that fermentation is not completely substituted, i.e. that the experimental state lies below the E.P. The proof of this will be that the R.Q. value shall be greater than 1·0. For this purpose the CO_2 output (our TR-res.) must be measured in association with the O_2 intake. For our present outlook the intake of oxygen is a strict measure of the true OR process, and its value enables us to evaluate the OR component of the mixed TR output.

What we require then from these two experiments in order to start our computation of S and of all dependent values of substitution is a set of three values: (1) NR-max., (2) TR-res., and (3) the R.Q. ratio. Three examples of the computation sequence are set out in table 1, but we may first indicate the theoretical lines along which we arrive at the values of S and of $OR_{E.P.}$ which are our objectives.

First stage. From the definition of R.Q. as being $\dfrac{OR + NR}{OR}$ and of TR-res. as equal OR-res. + NR-res., we can evaluate OR as equal $\dfrac{TR}{RQ}$ and hence arrive also at NR-res.

---

* The reference is to a paper by Dr D. J. Watson which has not yet been published.

Second stage. From the definition of TR-def. as NR-max. less TR-res. and the expression $S - 1 = \dfrac{\text{TR-def.}}{\text{OR}}$ (expounded in Paper VII, p. 144) we can derive $S - 1$ and hence S.

If we now turn to the headings of the nine columns in table 1 we can note the sequence of the computation. Columns 1, 2, and 3 contain the three observed values, NR-max. in nitrogen, and TR-res. with R.Q., in low oxygen concentration, which are essential for the start. The values of OR-res. in column 4 are obtained from column 2/column 3; and values of NR-res. in column 5 from column 2 minus column 4. Then for the second stage the values of TR-def. in column 6 are obtained from column 1 less column 2, while $(S - 1)$ in column 7 is derived from column 6/column 4. The values of S in column 8 emerge directly from the values in column 7. Finally, we can arrive at the critical values of $OR_{\text{E.P.}}$ set out in column 9 by the relation column 1/column 8.

Table 1

1 observed NR-max.	2 observed TR-res.	3 observed R.Q.	4 col. 2/ col. 3 OR-res.	5 col. 2 less col. 4 NR-res.	6 col. 1 less col. 2 TR-def.	7 col. 6/ col. 4 $S - 1$	8 from 7 S	9 col. 1/ col 8 $OR_{\text{E.P.}}$
10·00	9·33	9·33	1·0	8·33	0·66	0·66	1·66	6·00
	8·66	4·33	2·0	6·66	2·00	0·66	1·66	
	7·33	1·83	4·0	3·33	2·66	0·66	1·66	
5·00	5·00	5·00	1·0	4·00	0·00	0·00	1·00	5·00
	5·00	2·50	2·0	3·00	0·00	0·00	1·00	
	5·00	1·25	4·0	1·00	0·00	0·00	1·00	
3·00	3·66	3·66	1·0	2·66	− 0·66	− 0·66	+ 0·33	9·00
	4·32	2·16	2·0	2·32	− 1·32	− 0·66	+ 0·33	
	5·64	1·41	4·0	1·64	− 2·64	− 0·66	+ 0·33	

The greater part of the table is filled with illustrative data showing how these values would work out for a wide range of tissue types. It is of course to be understood that all these values are not drawn from any experiments but are illustrative, and calculated backwards for selected values of S to give a wide range of contrasts. The values are in three horizontal compartments, of which the first deals with relations for a tissue with the high S value of 1·66. The middle compartment exhibits a tissue with $S = 1·00$, while the lowest compartment exhibits a very low value $S = 0·33$.

Each compartment contains three lines of values which were calculated actually by starting with three OR values of 1·0, 2·0 and 4·0 respectively. But the set of three lines aims at presenting a sequence of three independent experiments carried out in three different low concentrations of oxygen (of unknown actual oxygen content). The three lines are arranged in order of increasing oxygen content as is shown by the marked fall in the R.Q. values for each group of three lines as the fermentation NR declines, and the series approaches the E.P. state where the R.Q. would show a value of 1·00.

It should be, of course, clear that all three of these lines of data are not required for one evaluation of S; any one line would suffice. The contents of a compartment have the form of an elaborate experiment carried out in triplicate, using three different oxygen concentrations to provide the observed data in columns 2 and 3, while the datum in nitrogen in column 1 is invariable, ready to be associated with any one of the three lines of data.

The next theoretical point that should be stressed is that the accuracy of the S values computed will depend upon the tissues used being throughout in the same state of carbon grade and carbon activation grade. The changes that we set out to measure in these data are changes of oxidation grade only. Only such data will give a clean measure of S, and any difference of carbon grade in the two coupled experiments will distort the computed S value.

We have also to stress the point that these computations tell us nothing about the location of E.P. (or any other functional locus) along an abscissal O_2 axis. What we learn in these calculations about $OR_{E.P.}$ is merely its pitch in terms of the ordinate axis. To decide its locus in relation to the other axis would require a series of additional experiments. This matter is of course of profound biological significance, but in this section we are considering where all the points concerned in substitution lie upon the ordinate axis only.

Finally, we may for a moment compare the contents of the three compartments as illustrative of substitutional behaviour in association with three such different S values. The essential differences appear in column 6 which sets out the TR-def. values. With high S these are positive as the TR slope declines from its initial value, while with S below unity the TR slope is a rising series and the TR 'deficits' are thus negative. With $S = 1.00$ the TR-res. values are constant throughout substitution, that is, in all values of what we call here *low* oxygen concentration. Consequently the TR-def. is zero throughout.

This section contains merely a formal exposition of the ideal relations of substitution with widely different S values. In the following paper* there will be presented a study of O_2 intake and R.Q. values for stored apples introducing the actual biological complications that develop.

2. The Inflected Forms of the Respiration Curves and their Variations

Graphic analysis of the (TR→OR) *inflexion.* The outstanding feature of the curves that we have established for the CO_2 production of apples, plotted graphically over an abscissa axis of oxygen concentration, has been the strongly marked break in form in the region of 5–10 % [O_2]. Here the graph reveals two convergent sloping lines forming a V-shaped depression with its trough over the locus of E.P. on the abscissa axis. In respiratory terminology this is due to the TR curves starting at a high value in nitrogen and continuing as a downward TR slope which breaks off at the TR-min. ($= OR_{E.P.}$) to become an upward inflexion, where TR is replaced by the rising CO_2 values of the OR slope, which slope may continue rising on to the OR values in air or even to 100 % O_2. Such a marked (TR→OR)

* See footnote p. 162.

inflexion form had not obtruded itself before for any respiration curve, and this feature became a starting point for an elaborate analytic inquiry into its interpretation, and for a comparison with the forms of other respiration curves.

In pursuit of our analysis of this inflexion feature we shall first make a brief analysis in graphic terms, and then take up the metabolic complexus behind this feature, in search of the precise biological determinants of the graphic form.

The rising and falling slopes in the graphs are generally somewhat curvilinear, but for first analysis we shall treat them as strictly rectilinear within the region of the inflexion. Their characteristic features will consist of starting points, endpoints, and intervening straight slopes. Variations of such forms can be stated in terms of the co-ordinates for graphic analysis, i.e. in terms of differences of pitch for units of the ordinate axis and in terms of units of extension for the $[O_2]$ abscissa axis. The first phase, the TR down-slope, drops from NR-max. to $OR_{E.P.}$, and its steepness depends on the abscissal extension between zero O_2 and the locus of E.P. The second slope, that of OR rise, starts from the $OR_{E.P.}$ and rises towards OR value in air, its steepness also to be defined by the extension of O_2 axis required for some standard rise of OR.

One graphic feature of the (TR→OR) inflexion that has practical importance is what may be called its 'obviousness', which depends primarily upon the magnitude of the angle of inflexion. Inflexion is very striking when the angle is about 90°, but as this angle widens out towards 180° the feature becomes less and less obvious, and will entirely disappear at 180°, being replaced by a straight line through the E.P. region although the underlying metabolism might remain of the same type as it was with marked inflexion. Another dimension that should contribute to the obviousness of the inflexion is the magnitude of the extension that it covers, along the abscissa axis. Were all the features of down-slope and up-slope run through within an abscissal extension of 1 % $[O_2]$, the phenomenon would have little chance of being characterized by experimental observations.

(a) *Examination of the metabolism in search of primary biological determinants of inflexion form*

Something has already been said in the previous section about *S* as a biological determinant in respiratory relations, but before taking up that thread again, we must examine the two fundamental aspects of the metabolism, fermentation and oxidation, for their contributions to the determinants controlling inflexion features.

Examining the function *F* in normal tissues of higher plants we find no cases recorded of the yeast type, in which fermentation takes place in air. We conclude therefore that the totality of the fermentation products can be put *in deficit* by the plant's aerobic respiration. This shows that the mechanism for the aerobic oxidation is present in excess and that it is the magnitude of the *F* function which acts as limiting factor to the amount of possible oxidative substitution.

The full measure of the *F* function is determinable, in nitrogen, by the value of NR-max. We therefore register that for the *F* function, this single point value, NR-max., is the primary biological determinant of the make-up of the inflexion feature, and as far as we know this is a real fundamental independent variable.

Turning now to the R function it may come first into our minds that there is a value in the category of OR which presents some analogy with the NR-max. of the F function, in that it is the corresponding value of OR used in substitution, namely, the value $OR_{E.P.}$. On second thoughts we recall that $OR_{E.P.}$ cannot have any of the elements of a determinant and an independent variable, for its magnitude is, on the contrary, under the control of the determinant NR-max., and is precisely defined by our critical expression $OR_{E.P.} = NR\text{-max.}/S$. It is then clear that $OR_{E.P.}$ is, as it were, a chance OR point in the sequence of the rising OR slope, which point happens to have the magnitude of $\dfrac{NR\text{-max.}}{S}$.

For the inflexion features, the aspect of OR that counts is the precise locus where this $OR_{E.P.}$ point comes in the graph. Where this may be found is determined graphically by the *steepness of the* OR *slope*, and it is the steepness of this rectilinear slope which is the true biological determinant on the OR side.

Steepness is an affair of co-ordinates, and its differences may be expressed as variation of abscissal extension to attain a defined pitch, or variations of pitch at a defined extension along the abscissa axis. For our special case of defining the steepness of varieties of OR slopes where we are considering an $OR_{E.P.}$ point of predetermined pitch, we must clearly make use of the variable abscissal extension required to attain this OR value. So we come back to measuring the axial extension from the origin in zero O_2 up to the locus of E.P. which identifies $OR_{E.P.}$.

The steepness that concerns the form of the inflexion on the graph is the steepness at the inflexion point E.P. and the adjacent rising course of the OR curve. This part of the OR curve is open to direct measurement and so presents no complications of interpretation. But the steepness of the OR slope from the origin up to E.P. is masked from direct observation. If it runs a strictly rectilinear course it is easily arrived at graphically, but if at all curved this part of the OR slope can only be arrived at indirectly by making certain postulates about the mechanism of oxidative substitution in the way that we developed this situation in § 1 of the previous paper.

When we proceed to inquire into the fundamental causation of variations of steepness of slope for the OR function we encounter at least two important and independent causes. There is first the efficiency of the oxidative and oxygen-reactive catalytic mechanism provided in the tissues for the utilization of the supply of molecular oxygen, and secondly the efficiency of the diffusive structure of these tissues in opposing the minimal diffusion resistance against the penetration of oxygen from the external atmosphere.

For the purpose of our present analytic survey we do not propose to go behind the combined effect of these two causes in giving a graphic appearance of varied steepness of slope to the OR curves but are content to bring forward as the second biological determinant and independent variable, the OR slope.

We may now recall our study of S values carried out in the previous section where it became clear that though the magnitudes of S depend upon underlying F and R metabolism, it is legitimate, at the present depth of analysis, to bring forward S as an independent variable and a biological determinant of inflexion form. We may regard it as a minor determinant when compared with the determinants from the F and R functions but entitled to a place in this heterogeneous assembly of which one member is a point-value, another an efficiency slope, and the third a functional ratio.

A superficial glance at the inflexion features of the (TR-OR) curve would suggest that if the OR up-slope figures as a determinant then the TR down-slope should also act in the same way. We propose therefore to hold a special inquest upon the nature of the TR slope to justify its exclusion from the physiological determinants. The TR slope may be defined by its co-ordinates, i.e. the ordinate drop from TR initial to $OR_{E.P.}$ and the extension horizontally from the origin to E.P. The drop is then from the value of NR-max. to the value of $OR_{E.P.}$; but as this latter has the magnitude of NR-max./S, the amount of the drop can be expressed as NR-max. $(1 - 1/S)$. The ordinate drop then is strictly determined by the two recognized biological determinants NR-max. and S.

When we turn to the extension that the drop covers we note that this is the extension that we adopted as an index of steepness of the OR slope—our other biological determinant. This short inquiry will serve to satisfy us that there is no element of an independent variable about TR slope but that it is a passive feature, strictly determined and not an active determinant.

For the sake of completeness we may insert here a paragraph to state that all this inquiry into the nature of the TR slope holds equally well for the cases where S is less than 1·0 and the empirical TR slope is a rising set of values (see § 1, pp. 160–1). With this type of rising TR the drop in pitch must be negative; but, also, just in this case, the defining magnitude $(1 - 1/S)$ will also be negative, since S is here less than unity.

As the TR slopes, though complex in their determination, are the one thing open to direct observation and measurement, it should prove worth while to have examined their nature fairly closely.

We are satisfied then that we have established that the biological determinants of the inflected form must be viewed as a triad with the striking contrast that for the F function the initial value counts but not the slope, while for the R function the OR slope counts but the initial value, being zero invariably for all plants, does not count as a determinant.

(b) *Diversity of inflexion forms produced by variation of the biological determinants*

In this section we propose to take the three biological determinants—OR slope, NR-max. and S—singly or in groups, and illustrate the diversity of forms of inflected respiration curves that may result through their variation.

One of the major differences that may result is variation in the magnitude of the inflexion angle and the obviousness of the inflexion. Before considering the more complex cases where both the OR and the TR side of the inflexion are undergoing variations we may illlustrate one case each in which the variation is limited, in the first case to the TR side and in the second case to the OR side.

Fig. 3 illustrates a set of variations on the TR side involving both its determinants, i.e. NR-max. and S, while the OR slope is taken as remaining constant throughout. There are shown, in all, nine TR slopes, three falling, three level and three rising, all impinging upon the one OR slope. These happen to give between them five loci of E.P. which are indicated upon the O_2 axis. The range of NR-max. illustrated comprises values 5·0, 10·0 and 20·0, while each is fitted with TR curves based on the S values of 0·5, 1·0 and 1·5. The resulting $OR_{E.P.}$ values that follow are defined by NR-max./S. It will be seen that a definite geometric inflexion angle is associated with each value of S. With $S = 1·5$ we have the angle of 90°,

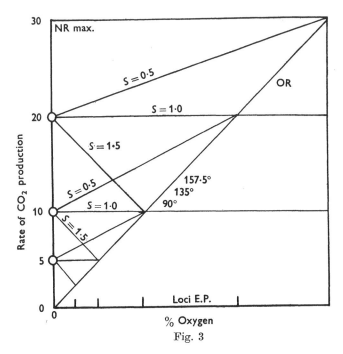

Fig. 3

with $S = 1·0$ the angle of 135°, while with $S = 0·5$ the angle is about 157·5°. The NR-max. and S values have been so selected that an identical $OR_{E.P.}$ value of 10·0 on the OR slope provides an inflexion point for three TR slopes, one from each of the three NR-max. initials. It is made clear that as NR-max. and S are both reduced in value, so the inflexion angle approaches the limiting value of 180° where any sign of inflexion would be completely eliminated, showing only a straight line through the $OR_{E.P.}$ value.

We now pass to illustrate the effect of variations in the determinant represented by the steepness of OR slope upon the form of the inflexion and especially upon the magnitude of the inflexion angle. For this purpose we might have drawn a set of OR slopes of different steepness, but more close contact will be made with the actual respiration of the apple if we obtain a series of less and less steep OR slopes all on one curve by adopting a hyperbolic form of the OR curve instead of a rectilinear. With such a curve there may be no two regions in which the slope is identical, but instead, a steady decline in steepness throughout. In a graphic figure we can arrange a series of inflexions distributed widely along the OR curve and see the variation of the angle of inflexion according to the local slope of OR at the inflexion point.

Fig. 4 is constructed to give a twofold illustration of the effect of declining slope along the median OR curve. The upper half of the diagram illustrates the results upon inflexion form when S is large, here 1·5, while the lower half gives the case when S is small, here 0·5. In this figure the graphic demonstration proceeds on special lines appropriate to a curved rather than a straight OR line. Associated with the S value of 1·5 we present four high NR-max. values A, B, C and D having the respective NR ordinate values of 20, 25, 28 and 30.

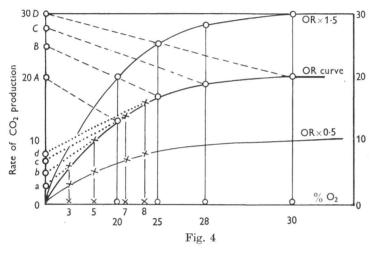

Fig. 4

Fig. 4. Four circles just above O_2 axis give the four loci of E.P. for values of $S = 1·5$ and N.R.-max. $= 20$, 25, 28 and 30. Four crosses just above O_2 axis give the four loci of E.P. for values of $S = 0·5$ and NR-max. $= 3$, 5, 7 and 8. Numerals below O_2 axis represent NR-max. values.

We proceed to draw above OR a second curve having throughout the ordinate values of OR $\times$ S and then draw horizontal lines from A–D to impinge upon this upper curve. The four contacts will give the vertical lines of the loci of E.P. for the four cases. Circles are here drawn upon the OR $\times$ S curve and perpendiculars dropped which cut through the OR curve and reach the $[O_2]$ axis. The four circles on the OR curve will give the four values of $OR_{E.P.}$, and those on the base-line the four loci of E.P. The actual TR slopes for the four cases will of course start at A, B, C and D, and decline, in curved slope, to end at the four $OR_{E.P.}$ circles on the OR curve, as delineated by broken lines. We see clearly in this figure how the angles of inflexion widen out as OR slope gets less and less at the successive loci of inflexion. Throughout the group, S is constant, and the mere rise of NR-max., though it makes the drop of TR greater from the beginning to the end at $OR_{E.P.}$, will not make the TR slope any less steep, as may be seen from the previous figure. The opening of the angle is directly due to the lessening OR slope.

It may be noted, for the series here figured, that if NR-max. should have happened to increase to a value above D then it would have become larger than the maximal possible value of OR × S for this drawing. NR would then never be extinguishable in any concentration of oxygen and fermentation would persist throughout.

We may now turn our attention to the lower region of fig. 4, and repeat our type of construction with $S = 0.5$. Here the OR × S curve lies below the OR curve. For this low S value we choose also low NR-max. values a, b, c and d (3, 5, 7 and 8 NR). For this set four crosses are drawn on the OR × S curve at the ordinate values equal to a, b, c and d respectively. Vertical lines through these crosses will cut the OR curve (above) giving the four $OR_{E.P.}$ values, and also cut the $[O_2]$ axis (below) giving the four loci of E.P. These points are all marked with crosses, and the four on the OR curve are joined with a, b, c and d to delineate the TR slopes which, for this low S set, are rising slopes.

With this set of rising TR slopes the inflexion angles are wider than with the high S set, and cases c and d come very close to attaining inflexion angles of 180°. Mere inspection of a graphic presentation of data could hardly establish the presence of an inflexion in such cases.

The conditions for elimination of inflexion at an angle of 180°. We may now glance at the conditions which would eliminate any sign of (TR-OR) inflexion, so that the line of CO_2 production above E.P. continued exactly the line of CO_2 production below E.P. It is of course only when S is very small and TR shows itself as a rising slope that there is approximation to this condition. Fig. 5 illustrates such a case where NR-max. is 3·3 and $OR_{E.P.}$ is 10·0, and therefore $S = 0.33$. The line of TR is shown continued, broken, beyond E.P. Any cause which lowered the slope of OR beyond E.P. so that its slope coincided with the dotted line would eliminate all sign of inflexion as a mark of the locus of E.P.

The slopes of the three lines which enter into the geometry at the locus of E.P. can be expressed in ordinate units as a rise or fall over the extension from the zero to E.P. on the abscissal axis. Or they can also be related in terms of S so that we can give a general expression to the change required to eliminate inflexion. For fig. 5:

In ordinate units:

TR rises $+6\cdot6$, OR rises $+10\cdot0$, NR-res. falls $-3\cdot3$.

On S ratios:

TR rises $+(1-S)OR$, OR rises $+1\cdot0(OR)$, NR-res. falls $-S(OR)$.

Whence if OR slope is depressed from 10·0 to 6·6, inflexion is eliminated. Or if OR slope is depressed from $(1\cdot0)(OR)$ to $(1-S)OR$, inflexion is eliminated.

We can generalize this last line into the statement that any intrinsic change in efficiency of OR function which lowers its efficiency slope over the $[O_2]$ axis, from 1·0 down to $1-S$ will eliminate all evidence of inflexion. The higher the S value the lower $1-S$ will be and the greater the depressant of OR must be to eliminate inflexion.

The appearance of rapid intrinsic lowering of the OR efficiency might be associated with the OR curve showing a primitive limiting-factor form or with rather abrupt flattening of slope in the stage by stage rise of an OR curve of the

enzymatic hyperbola form. We may add a pair of figures showing possible interactions of oxidative substitution with an OR curve showing the primitive limiting
factor inflexion in which the rising initial straight line passes over abruptly into
the horizontal maximal rate of OR.

In fig. 6, S is illustrated as having the value 0·40, so that taking $OR_{E.P.}$ as 1·0,
NR-max. is 0·40. Here E.P. is shown as identical in locus with the intrinsic attainment by OR of its maximal value and horizontal graphic line. There is here no

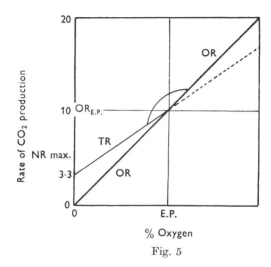

Fig. 5

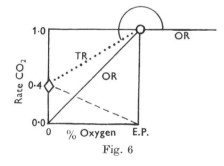

Fig. 6

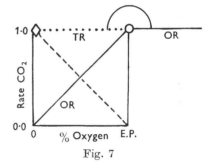

Fig. 7

upward inflexion in the CO_2 curve, at E.P.; and, indeed, as TR passes over into OR
there is an angle of well over 180° which might be described as a downward inflexion.

In fig. 7, S is taken as equal to 1·0 so that NR-max. and $OR_{E.P.}$ are identical, and
we arrive at a CO_2 curve which holds the value of 1·0 right from nitrogen on to
high $[O_2]$ with no trace of angular inflexion at E.P.

We may append to this consideration of the elimination of appearances of
inflexion a reference to another possibility which may cause an inflexion feature to
be overlooked in experimental work.

Obscuration of an inflexion when E.P. *is located close to the zero value of* $[O_2]$. In
apples the E.P. is located in the range of $[O_2]$ between 5 and 10 %, a region where
it is easy to provide a set of controlled oxygen concentrations for experimental
exploration of the angle of inflexion. It is conceivable, however, that with very

low values of NR-max. and very steep rise of the OR slope, the full value of $OR_{E.P.}$ would be obtained even in say 0.25% external concentration of oxygen. In such a case it would be very difficult to prove that there was a TR-min. at precisely that concentration. The whole phenomenon of substitution would be telescoped up near the origin on the O_2 axis and might well defy practical experimental analysis or even identification. So we conclude that it may remain difficult to establish the presence or absence of our schematic oxidative substitution relations in tissues of high efficiency in utilizing external oxygen, and that on the contrary sluggish tissues of high diffusion resistance are more suitable for exploring the theoretic rules of this substitution.

We may close this exposition by summing up the physiological characteristics which tend to obscure the inflexion feature in respiration curves. These are low NR-max., low S and steep OR slope, which latter implies high oxidative efficiency and low diffusion resistance.

(c) *The ontogenetic drift of biological determinants and the sequence of inflexion forms*

In working through the analysis of inflexion forms in the previous sections the object we had in view, beyond that of establishing the influence of the three biological determinants, was a survey of the divergent types that the respiration of other plants than the apple might present. We examined diversity of inflexion features in the widest biological sense, setting out possible cases the occurrence of which is not yet established or suspected.

There is, however, a second survey of diversity of inflexion that has now to be taken up. This is of much more limited range and is confined to the respiration of apple tissue. Its justification lies in our need to unravel the actual changes of inflexion form that the respiration curves exhibit as the tissues gradually develop in storage through their ontogenetic stages.

This phenomenon is at once presented to us by the diversity, in detail, of the four inflected curves that are called for to present the respiration data in our main collective figure (see fig. 1 in Paper VII). It is not easy to carry out a process of analysis directly upon this set of four curves. We propose therefore to make use of the principles that we have enunciated in recent sections and to construct, by aid of these, a variety of charts of drifting inflexion forms of known definition. Inspection of these synthetic charts will enable us to decide which particular one exhibits the features of the actual apple data. In this way we may hope to close in upon the drifts of magnitude of the biological determinants that are at work underlying the ontogeny of apple tissues.

The determinants we have already decided to be three in number, with, as primary ones, the OR slope and the NR-max., while S is to be added as a secondary determinant.

We may now turn to describe the graphic machinery of construction of the series of charts, presented in figs. 8 and 9, that will serve as models for the effects of the variable determinants upon inflexion form. The abscissa axis in all cases represents an arbitrary scale of $[O_2]$, the ordinates are CO_2 values; in all cases four ordinate values are marked to the left as a rising series a, b, c and d, and four to the right as A, B, C and D. The a, b, c and d series stands for the NR-max. variation range, while ruled lines rising from zero to join the four A, B, C and D values would set out the variation range of the four rising OR slopes. Should the OR slope be postulated as invariable in any particular chart then only one of the values to the right (arbitrarily, always C) is selected to be furnished with a ruled OR slope. Similarly, should NR-max. be postulated, in its turn, as invariable, then only the value c is used as the origin of declining TR slope.

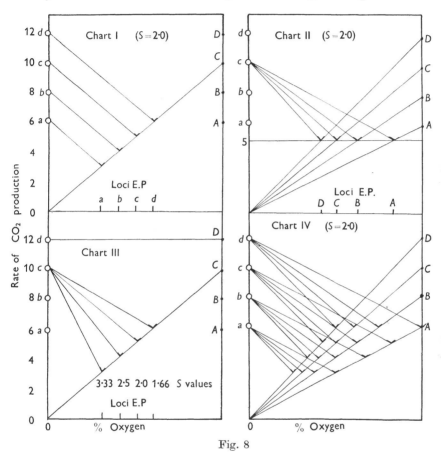

Fig. 8

When S is held to be constant throughout a chart, its value is always taken as $S = 2\cdot 0$. If S is variable, the values adopted, ranging on either side of $2\cdot 0$, are written in the chart against the TR slopes.

With this postulation of the NR-max. values and of the S values the pitch of $OR_{E.P.}$ proper to each TR curve is at once arrived at, and can be marked on the associated OR slopes. This locus will constitute the inflexion point where the falling TR slope from NR-max. will pass over into the rising OR slope. These inflexion points are stressed by drawing heavier local lines which indicate the inflexion angle.

Lines dropped from these inflexion points on to the abscissa axis will give the locus of E.P., in terms of the arbitrary scale, in x units, of oxygen concentration.

Eight charts are to be presented altogether of which the first three will represent what we may call: *Variations of the First Order*, and the last five: *Variations of the Second Order*. Within the first order we include charts which represent variation of only one single determinant out of the three that have just been formulated. Within the second order are included those in which the two primary biological determinants, namely, OR slope and NR-max., are both undergoing, simultaneously, independent drifts of variation.

Chart I of the first order. This illustrates the appearance produced by four steps of variation in the value of NR-max. upon the inflexion form. Here the OR slope is shown as being devoid of change or ontogenetic drift, maintaining constantly

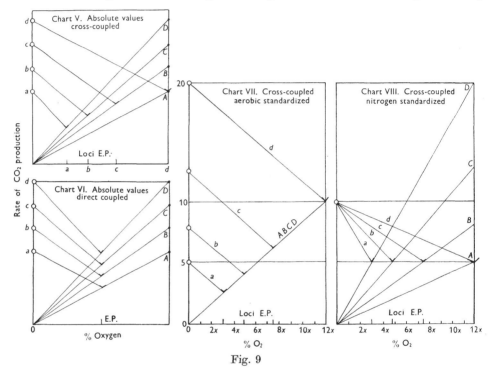

Fig. 9

the slope characterized as *C* while NR-max. drifts through a series. As *S* is also constant (given here the value 2·0), then with drift of NR-max. from 6·0 to 12·0 the $OR_{E.P.}$ values must drift from 3·0 to 6·0. These four $OR_{E.P.}$ values are shown as a drifting series up the OR slope. As a concomitant of their displacement, the locus of E.P. will drift away from the origin through increasing values along the $[O_2]$ axis. It will be noted that for this type of drift, though the extension of the inflexion increases, the inflexion angle remains constant, provided of course that the OR slope is strictly rectilinear over the course under consideration.

Chart II. In this chart, NR-max. is invariable, while OR slope is shown as the drifting variable; all four variants of it, the slopes *A*, *B*, *C* and *D*, being drawn in. On the NR-max. side only the point *C*, with value 10·0, is involved. As $S = 2·0$ throughout, $OR_{E.P.}$ must be 5·0 on all four OR curves, marking the four inflexion points. As OR slopes drift from *D* to *A* this chart demonstrates that loci of E.P.

drift away from the origin with accelerating pace, and as this progresses the inflexion angles get wider and wider.

Chart III. This third representative of the first order of variation shows the effect upon inflexions when S provides the only variable and both NR-max., at the value c, and OR at slope C, are invariant throughout the drift of S. For variants of the S value we have an arbitrary choice and we present values varying either side of the value of 2·0, the constant value of the other charts. The series of values here exhibited is the set that will give the same scatter of loci of E.P. as we arrived at in chart I. The values required are 3·33, 2·5, 2·0 and 1·66, written beneath the $OR_{E.P.}$ values of 3·0, 4·0, 5·0 and 6·0.

In this chart then the $OR_{E.P.}$ values drift along the OR slope and the angles widen as S diminishes.

Chart IV: Introduction to the second order. Our study of the ontogenetic drift of apple curves leads us to believe that all the three charts of the first order are too simple to interpret the whole of the happenings in apples in storage. We must then examine the cases in which both NR-max. and OR slope are undergoing simultaneous independent drifts.

The simplest approach to selected charts of the second order seems to be through chart IV, which contains a formal assembly of the maximum of combinations of variation that our four OR slopes and four NR-max. values permit, provided S remains constant. Here S is still taken as 2·0. It hardly needs any further exposition to demonstrate that there will result four values of $OR_{E.P.}$, and that all four could be represented on each OR slope. We are thus provided with a set of sixteen inflexion forms showing a wide scatter of E.P. locus along the $[O_2]$ axis. All these sixteen curves can be identified by labelling them with the sixteen combinations of a, b, c, d with A, B, C, D. From this comprehensive assembly we can pick out sets of four curves to make up a chart to illustrate the point that we want to stress in dealing with the second order of variation.

Charts V and VI, fig. 9. In chart V we have collected the four inflected curves aD, bC, cB and dA while in chart VI we have aA, bB, cC and dD. Both show the same upward series of NR-max. values, but in chart V this series is coupled with a falling series of OR slopes, while in chart VI the coupling is to a rising series of OR slopes. It is clear that when we draw up charts with both NR-max. and OR slope varying, there are two types of coupling. Either there is *cross-coupling* of the independent variables so that while one drifts to higher values with ontogeny the other drifts to lower values. This is the coupling in chart V, while in chart VI we have the opposite type that we may call *direct coupling* where both independent variables are drifting in the same sense.

The general appearance of the two charts is very different as regards drift of loci of E.P. and angle of inflexion, though the pitch range of the four values of $OR_{E.P.}$ is the same for both. The differentiating geometrical feature is that in chart V the highest TR slope cuts across all the OR slopes, while it is the lowest TR curve that cuts across the OR slopes in chart VI. This is a character which should not be obscured in empirical charts, even if the OR slopes exhibited some curvature. If we inspect our empirical family of ontogenetic curves for apples (fig. 1

in Paper VII), it should be obvious that the apple belongs to the type which we label cross-coupled, so that with ontogeny, NR-max. and OR slope are drifting in opposite senses as independent variables.

Granting that the apple is established as an example of cross-drifts, we have yet two possibilities in relation to actual time sequence. The apple illustrates the alternative that, with senescence, NR-max. rises while OR slope flattens; and not the case that OR slope steepens while NR-max. falls.

The distortions due to standardization. Charts VII and VIII. All this survey of TR inflexion forms has been presented so far as if we were dealing with direct *absolute* values of CO_2 production for our ordinate values, both for NR-max. and for OR slope.

We must, however, recall, as an appendix to this presentation of charts in absolute values, that the determinations of CO_2 productions in this particular research on apples have been presented in all the essential graphs as relative values, typically relative to an 'air' standard. All the CO_2 values are then really ratios to CO_2 production in air taken as unity; strictly they are all OR/OR_A values, as we have stressed before (see Paper VII, p. 145).

It is therefore important that we should make clear what would be the effect upon these charts of inflexion forms, if we took an absolute chart such as chart V and transposed it into a relative chart, either to the 'air-standard', as is usual, or, for instruction, into the 'nitrogen-standard', where all CO_2 is given as a ratio to NR-max. observed in nitrogen as unity.

The final pair of charts presents the inflexions of chart V in these two trans-positions. In chart VII (fig. 9) the critical values of chart V have been recalculated on the basis that the CO_2 production in 'air' is for each curve to be taken as 10·0 units. Table 2 sets out the 'absolute' values of charts V and VI and also the values of V transposed to the air and the nitrogen standards respectively as VII and VIII. These synthetic charts have only an imaginary oxygen concentration axis in x units, so there is no fixed location for air; we have taken the ultimate point on the abscissa axis labelled $12x \% O_2$ as equally significant as deputy for an aerobic standard for the moment. The aerobic ordinate values over this point are in chart V 6·0, 8·0, 10·0, 12·0. Bringing all these to the value of 10·0 we get one common OR slope for all four curves in chart VII. On this are stretched out the four new adjusted $OR_{E.P.}$ values 2·5, 4·0, 6·25 and 10 due to the NR-max. values having now become 5·0, 8·0, 12·5 and 20·0.

In the companion chart VIII, the data of V have been transposed to the nitrogen standard where all four NR-max. values have been standardized to 10·0 and all $OR_{E.P.}$ values consequently to 5·0. This carries with it the new series of values for OR over the $12x \% O_2$ locus of 5·0, 8·0, 12·5 and 20, so that the OR slopes diverge on this standard very widely.

In all three charts the loci of E.P. on the arbitrary $[O_2]$ axis will be identical, namely, at 3·0, 4·75, 7·4 and 12 units of $x \% O_2$. This objective feature of the relation of the locus of the inflexion point to the external oxygen concentration has, of course, not been shifted by our standardization manipulation of the relative values of ordinates.

The conclusion of this section lies in the comparison of these last two model synthetic charts with the already presented empirical apple results on the air and nitrogen standards respectively (see figs. 5 and 6 and § 2 (*f*) of Paper VII). It seems clear that the actual figures given there are of the type worked up in these two models. This treatment supports the position taken on other lines of evidence that with increasing ontogenetic development, the OR function declines and the NR function rises in relation to it. The exact distribution of the relative contrary movement between the two functions must await the survey of absolute data to be brought forward in a later paper.

Table 2

Chart VI. Direct-coupled curves in absolute values					Chart V. Cross-coupled curves in absolute values				
	NR-max.	$OR_{E.P.}$	OR at 'air'			NR-max.	$OR_{E.P.}$	OR at 'air'	
d	12·0	6·0	12·0	*D*	*d*	12·0	6·0	6·0	*A*
c	10·0	5·0	10·0	*C*	*c*	10·0	5·0	8·0	*B*
b	8·0	4·0	8·0	*B*	*b*	8·0	4·0	10·0	*C*
a	6·0	3·0	6·0	*A*	*a*	6·0	3·0	12·0	*D*
Chart VII. Cross-coupled, air standard					Chart VIII. Cross-coupled, nitrogen standard				
	NR-max.	$OR_{E.P.}$	OR at 'air'			NR-max.	$OR_{E.P.}$	OR at 'air'	
d	20·0	10·0	10·0	*A*	*d*	10·0	5·0	5·0	*A*
c	12·5	6·25	10·0	*B*	*c*	10·0	5·0	8·0	*B*
b	8·0	4·0	10·0	*C*	*b*	10·0	5·0	12·5	*C*
a	5·0	2·5	10·0	*D*	*a*	10·0	5·0	20·0	*D*

It will be noted that we have not developed the proposition that there is, during ontogeny, a drift of *S* values given by the third biological determinant. There is evidence presented by the empirical data pointing in this direction; in that the latest value of *S* is the highest, but the three earlier values show no evidence of drift. What would be the effect of a rising drift of *S* can be seen by the illustrative chart III in fig. 8.

3. THE CARBON METABOLIC SETTING OF THE FUNCTIONS OF RESPIRA-
TION AND OXIDATIVE SUBSTITUTION IN LOW CONCENTRATIONS OF
OXYGEN

(*a*) *The interrelations of activation of carbon and*
oxidation by oxygen

In our preceding study of the relations of fermentation and respiration we have concentrated attention upon the phenomenon of oxidative substitution, and we have endeavoured to treat this aspect of respiratory metabolism as an abstracted isolated happening of which the main interest was the ratio of substitution —*S*—.

In living tissues this substituting system is immersed in other metabolic activities which can distort its working. The aspects of some of these effects may be evaded, but certain parts of the associated metabolism are essential for producing the carbon reactants involved in fermentation and respiration. These latter can no longer be evaded, and we must now face them and characterize them, formulating what we propose to call the carbon metabolic setting of the oxidative reaction.

At once we recognize that we have to manipulate a mixture of determinants of rate, some concerned with the carbon function and others with the oxygen function, the two sets interacting to make up the intermediate metabolism of fermentation plus respiration.

We propose to introduce special terminology for the aspects that appear to us to be of fundamental significance. On the carbon function side we may begin with what we shall call *I, The Carbon Grade*. This indicates the grade of production rate at which carbon is being made available from reserve sources for intermediate metabolism. It might be thought of as the rate of production from insoluble substances of soluble sugars in general. There is a tendency for this rate to decline when plant organs are isolated from sources of supply and this decline is known in a general way as 'starvation'.

With apples, which are capable of such long independent existence, the grade of starvation shifts only very slowly, so that for short periods of time we find the appearance of static states of carbon equilibrium. Within the normal range of these carbon grades we get the impression that the ultimate carbon sources for F or for R are closely coupled, and we adopt the view that the carbon sources of OR and NR decline together in sympathetic dependence during starvation.

The next stage of carbon metabolism that we give prominence to we may call *II, The Carbon Activation Grade* (or, for short, carbactivation). This also is a rate— the rate of preparation of the specific substrates for the functions of F and of R. It is outside our present formal analysis to discuss whether these carbon substrates are special hexoses or trioses or compounds such as glycophosphates; we regard them simply as 'activated' carbon. One special complexity about this function is that the grade or rate of it appears to us to be heightened by the presence of oxygen.

This oxygen acceleration of the grade of carbactivation we regard as a process, which though it leads to an increase of OR production is essentially different from increase of oxidation rate by greater concentration of O_2 as a reactant in this oxidation.

The special activity of oxygen now under consideration is presented as due to a shift of efficiency of the carbactivation mechanism through its entering into a new adjusted state in equilibrium with the partial pressure of free oxygen in the cells of the tissue. The maintenance of this higher efficiency is not held to involve continued consumption of oxygen. In analogy with phenomena recognized in animal metabolism it might be held that the partial pressure of oxygen determines the proportion in which some accessory agent is distributed between its oxidized and reduced states; and that of these two states only the oxidized state functions as a co-body in the catalytic mechanism of carbactivation.

In carbactivation, as in carbon grade, we believe that there is a close linking up of the grade with both the NR and the OR functions; so that this rise of grade produced by oxygen would heighten the potentiality of carbon production for NR as much as the actuality of carbon production for OR. Special experiments in test of this thesis have yet to be prepared for publication.

The third category needed for measurement of the activity of the carbon function is one which makes contact with the oxidation function, in the narrowest sense, picturing oxygen as a reactant in the oxidation of carbon molecules for aerobic respiration.

We propose to call this measure, *III, The Oxidation Grade*, and use it to register the extent to which the active carbon produced by II is diverted from the fate of becoming fermentation products to the fate of aerobic respiration products. Our index of this diversion is the degree of oxidative substitution of OR for NR. As there must be an upper limit to this grade of oxidation reached when all the available carbon undergoes respiration and all the CO_2 given off has the label of OR, we may best express oxidation grade in percentages, ranging from 100 % as just indicated down to 0 % in nitrogen.

This definition of oxidation grade is really part of the general conception of oxidative substitution, but what we are now to concern ourselves with is not the value of the coefficient S which governs the NR/OR relation in this substitution, but the degree of completeness to which this substitution has been pushed by the external $[O_2]$ supplied to the tissues. When the oxidation grade rises to 100 % we arrive at the E.P., the critical locus where all NR has been extinguished and OR has risen to the value defined as $OR_{E.P.}$. Any further increase of external oxygen cannot increase the oxidation grade; though the accumulation of free oxygen, causing rising partial pressure of O_2 in the tissues, can increase the rate of the carbactivation grade and so send up the absolute amount of active carbon oxidized, while the oxidation grade continues indefinitely at 100 %.

We have next to consider the arithmetical complications that arise when we wish to compute the percentage oxidation grade for data of mixed OR + NR production in low $[O_2]$. We can only resolve an observed TR value in this condition into its components when we have, in addition, knowledge of two out of the three values. NR-max.—S—$OR_{E.P.}$ related by $\frac{\text{NR-max.}}{S} = OR_{E.P.}$ It is clear then that the observation in $[O_2]$ must be coupled with certain control observation for determination of these critical values. NR-max. could be arrived at by a coupled observation in nitrogen and $OR_{E.P.}$ by an observation at the E.P. locus.

An example may be given. Let the controls give us the values of NR-max. = 100, and $OR_{E.P.} = 66 \cdot 6$, whence $S = 1 \cdot 5$. If the observed TR value in low $[O_2]$ is $90 \cdot 0$ we set out in this state NR-res. as x and NR-def. = $100 - x$. We arrive at two expressions for the OR-res. component of TR; by $\frac{\text{NR-def.}}{S}$ we get $\frac{100-x}{1\cdot5}$, and by TR less NR-res. we get $90 - x$. By equating these $x = 70$, whence OR $= 20$, being 30 % of the maximum yield of $66 \cdot 6$. The oxidation grade in the examined low O_2 state is therefore 30 %, while 70 % of the carbactivation rate is left as the fermentation grade.

Several general expressions for the oxidation grade may be set out based on approach from the OR or NR side:

(1) $\dfrac{\text{OR-res.}}{\text{OR}_{\text{E.P.}}}$,

(2) $\dfrac{\text{OR-res.}}{\text{OR-res.} + \text{OR}_{\text{E.P.}}\text{-def}}$,

(3) $\dfrac{\text{OR-res.}}{\text{OR-res.} + \text{NR-res.}/S}$,

(4) $\dfrac{\text{NR-def.}}{\text{NR-max.}}$,

(5) $\dfrac{\text{NR-def.}}{\text{NR-def.} + \text{NR-res.}}$,

(6) $\dfrac{\text{NR-def.}}{\text{NR-def.} + \text{OR}_{\text{E.P.}}\text{-def.}/S}$.

All these expressions give the fractional oxidation grade, and $\times 100$ give the percentage oxidation. Their interrelations are easily seen from the substitution diagram of fig. 1.

We may now stress the final point that since estimation of oxidative substitution and oxidation grade demand coupled sets of experiments in different concentrations of oxygen, it is essential that all the members of the set shall have identical carb-activation grades. Or, if not identical grades, they must have grades differing in a well-established numerical relation. This is the key position for accurate evaluations of oxidation grade, and some close scrutiny must be given to this matter.

The essence of the matter, in experimental work, is to have both these possible augmentation effects of oxygen continually before one, and to provide that any observed rise of OR is assigned critically among the three grades of the carbon function. We may express the differentiation of functions that we have in mind by a segregating system of brackets and afterwards follow up this analysis by a graphic chart of the functional distinctions proposed for our actual apple data.

The schema is set out below. The determinants of grades I and II appear outside the square bracket, while the oxidation relations of III are collected inside this major bracket. This stresses that grades I and II have a common effect upon both of the included pair of curved brackets that express the range of substitution relations between OR and NR that proceeds antithetically under the control of oxygen as an oxidizing agent. The top lines within the curved brackets give the states at E.P. where NR is zero and OR is $\text{OR}_{\text{E.P.}}$ and the bottom lines the states in zero oxygen. These relations and the intermediate stages of substitution are all controlled in numerical ratio by the coefficient S.

Before leaving this schema we may stress the fact, which was much used in the earlier papers of this series, that the primary practical distinction between the working of the systems controlling grades I, II and III is that while oxidation changes of III take place rapidly in transitions the changes of grades II and I take place very slowly in apples, so that the mechanisms directly reveal their distinctions in time sequences after a change of oxygen conditions.

$$
\begin{array}{l}
\text{Control of carbon supply} \\[2ex]
\text{Factor I, carbon grade} \\[4ex]
\text{Factor II, carbactivation grade}
\end{array}
\left[
\begin{array}{c}
\text{III. Oxidation grades determined by oxidative} \\
\text{substitution:} \\[1ex]
\begin{pmatrix}
\text{OR}_{\text{E.P.}}\text{ at E.P.} \\
\text{OR-res.} \\
\text{OR-def.} \\
\text{OR}_N\text{ at zero}
\end{pmatrix}
\times S =
\begin{pmatrix}
\text{NR-zero at E.P.} \\
\text{NR-def.} \\
\text{NR-res.} \\
\text{NR-max. at zero}
\end{pmatrix} \\
\qquad\qquad \text{OR} \qquad\qquad\qquad\qquad \text{NR}
\end{array}
\right]
$$

(b) *The relation of carbactivation and oxidation to the* $[O_2]$ *axis*

Having stressed the separate individuality of the conceptions of oxidation and of carbactivation, we may now show their suggested application to the analysis of the whole range of the (TR→OR) curve for apple respiration.

Fig. 10 shows schematically, over one axis of $[O_2]$, three curves, of which the top presents the CO_2 values of the observed (TR→OR) curve, while the middle one stands for our carbon curve, that is it presents the assumed drift of the carbactivation factor in its relation to $[O_2]$. The lowest curve presents the curve of oxidation grade in percentage of the carbon possibility which is being determined by the carbon curve.

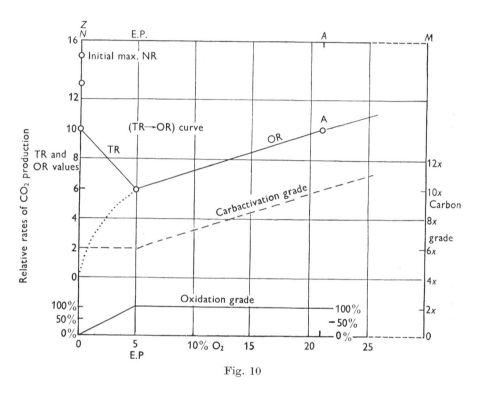

Fig. 10

In terms of this binary background we have to set out the drifting determination of respiratory magnitudes as the oxygen concentration rises along the abscissa axis. To simplify our terminology in this progress we attach to the TR-OR curve the five labels which we introduced in the previous paper (p. 139) as locative sign-posts of functional states along the O_2 axis. The significance of the locative suffixes, N, E.P. and A is clear, while M, standing for the maximal value at the optimal $[O_2]$, lies outside our present view. The symbol Z has a very special significance, as a higher value than the N value located in zero oxygen. The precise application will be expounded later when wanted (see p. 187).

The form of the (TR-OR) curve at the top of the figure with its well-marked inflexion at E.P. is very familiar, but the forms of the two component curves below it show special features which we must now discuss in some detail. The curious form of the carbon curve is held to arise out of the special nature of its activation by oxygen. It is held that this special effect of oxygen concentration depends upon the concentration of free oxygen in the tissues maintained by the residue after the high affinity of the oxidizable substrate for oxygen has been satisfied in oxidation.

This residual free concentration of O_2 is considered, as a first approximation, to be kept at zero through the range of low external concentrations of O_2 until the oxidative substitution is complete, that is, right up to the locus of the E.P. line. Beyond this locus, free oxygen begins to accumulate and the carbactivation grade rapidly rises. The graphic form shows a level grade of availability of carbon right from the initial TR (NR-max.) region on to the locus of the $OR_{E.P.}$ value. An arbitrary scale of carbactivation grades is added on the right, and we see that beyond E.P. this curve is given a form exactly proportional to that of the OR curve in all upper regions, showing also $6 \cdot 0x$ at the E.P. and $10 \cdot 0x$ in air. This sequence of carbon values could be distinguished by adopting for this carbon curve the same locative subscripts invented for the TR→OR curve. We should thus get C_N, $C_{E.P.}$ (these first two having in this case, identical values) followed by C_A and, later, C_M.

The oxidation grade curve, at the bottom of the figure shows a form which is quite different, rising from 0 % in nitrogen up to 100 % at E.P., where the oxidative substitution has completely extinguished all the NR component. Beyond E.P., O_2 supply ceases to be limiting for oxidation, and there is oxygen in excess of oxidative consumption, so that the oxidation grade continues at 100 % with every further rise of carbactivation. The oxidation grade, when expressed, in units of carbon, in order to rank with our other computations, can be formulated as $\dfrac{OR \times S}{OR \times S + NR} \times 100$. The three curves combined in this graph present our picture of respiratory control in apple tissues and bring out the critical nature of the E.P. for all functions, marking the change-over from oxidative control of TR values to the control by carbactivation as a limiting factor.

Ontogenetic drift of carbactivation and oxidation grade values. It has already been stressed that the locus of E.P. drifts along the $[O_2]$ axis away from the origin during the progress of ontogeny and senescence with apples. As E.P. locates the inflexion point of the form of the (TR→OR) curve this drift of E.P. was fundamental to our analysis of inflexion form in the last section. So also for the drift of the grades of oxidation and carbactivation, this shift of E.P. must be fundamental, as E.P. delimits the locus of change-over from control by factor III to control by factor II. From zero O_2 to the E.P. the carbactivation grade is held to remain at its low anaerobic level, while through this range the oxidation grade is rising to 100 %. In apples these relations bring about a decline of total TR production. Beyond E.P. the rise of CO_2 is due directly to rise of carbactivation.

As E.P. drifts onward with ontogeny all these phases will become more and more extended along the O_2 axis and therefore show a lower and lower pitch of grade for any external $[O_2]$ value used as a standard.

We must recall that the fundamental oxygen relation which decreases with ontogeny is the steepness of the OR slope which is the index of the tissue *efficiency* for the utilization of external oxygen, compounded of catalytic efficiency and diffusive efficiency. Decreasing efficiency pushes the locus for attainment of OR$_{E.P.}$, with its associated controls, further along the axis. The oxidation grade consequentially attains 100 % grade only further along the axis, and the efficiency of oxidative substitution as set out within the bracket on p. 180 is reduced. Consequentially the oxygen acceleration of carbactivation rate, as segregated outside the same bracket, is postponed till this 100 % oxidation grade has been attained.

(c) *Ratios along the* (TR-OR) *curve, and their local analytic significance*

Fig. 10 gives definiteness to the curve of CO$_2$ production in apples by asserting that this result is the outcome of two major components, the carbactivation grade and the oxidation grade. The relative activity of these two factors in control of rate of CO$_2$ production is so different in regions above and below the locus of E.P. that the analytic teaching of ratios of CO$_2$ production drawn from two different points on the curve will depend entirely upon where the two ratios are drawn from.

The ratio drawn from certain pairs of loci will illuminate the value of S, that from other pairs will give information about carbon grade, while others again will give values for oxidation grade.

We may say that every ratio drawn from a pair of CO$_2$ values taken at different [O$_2$] loci has elements of more than one of these controlling grades within it whether it is a ratio of NR/NR, or OR/OR or of NR/OR.

It should therefore clarify matters to survey as a whole the types of ratios that can be extracted from such a CO$_2$/[O$_2$] curve as the apple provides. In the earlier stages of analysis of the empirical ratios, our prime thought was to find ratios which would illuminate one single function free from distortion by another function, but later we came to recognize the universality of both carbon and oxygen functions and so tried to display each ratio as being inevitably composite. For this purpose we have formulated the duality of ratios with a carbon factor placed outside a bracket, while within the bracket is enclosed the oxidation factor. This graphic treatment was sketched on p. 180 and will be utilized in the present development of types of ratios. We may survey in a series of paragraphs different types or families of ratios drawn from the (TR-OR) curve in fig. 10.

The NR/OR *ratios selected for the evaluation of the substitution coefficient S.* In the earliest stages of our study of respiration (TR values) in low oxygen we groped for an evaluation of the exchange value of NR for OR when fermentation was put in deficit by the use of oxygen supply for oxidative substitution. We first arrived at such values of S by relating the initial TR value (adapted NR-max.) to the OR production at the TR-min. The more scientific formulation of this evaluation was to write $\dfrac{\text{NR-max.}}{\text{OR}_{E.P.}} = S$. This has been already fully developed; all we wish to do here is to draw up exactly what are the characteristics of an empirical CO$_2$ ratio of production which qualify it to be used for the calculation of S, which is in theory a relation between an NR-def. and an OR production. In each CO_2 value, i.e. both in the numerator and in the denominator of this expression, there must be a determinant element of carbon grade as well as one of oxidation grade. Calling these components C and O.G. respectively, the full make-up of the above ratio must be threefold:

$$\frac{\text{NR-max.}}{\text{OR}_{\text{E.P.}}} = \underset{\dfrac{C}{C}}{\overset{\substack{\text{Carbon} \\ \text{factor}}}{}} , \quad \underset{\dfrac{\text{OG}}{\text{OG}}}{\overset{\substack{\text{Oxidation} \\ \text{grade}}}{}} , \quad = \frac{S}{1}$$

These components are different in their arithmetical effect. Carbon grade is a rate of metabolic carbactivation, and therefore if the S value is not to be distorted, C in the numerator NR must be an identical magnitude with C in the denominator OR; but relative oxidation grades have a different form of expression, being percentages of the potential oxidizable carbon that is actually oxidized in the given $[O_2]$. For the numerator, in nitrogen, the O.G. value has to be 0%, and for the denominator in $[O_2]$ at the E.P. locus the O.G. value has to be 100%. Unless these conditions for the carbon function and the oxidation grade are satisfied then the derived S value will not be a pure value for the coefficient of oxidative substitution.

Should there be a trace of oxidation in the nitrogen or a slight residue of fermentation in the $\text{OR}_{\text{E.P.}}$ value so that the O.G. expression were $\dfrac{1 \cdot 0\%}{99 \cdot 0\%}$, then the S value would be slightly vitiated in a way that is arithmetically indirect but easily calculated. If, on the other hand, the carbon grade at the selected OR were really slightly above that appropriate to $\text{OR}_{\text{E.P.}}$, then also S would be distorted; in this case the departure of C/C from unity would affect the S value directly and proportionately.

Finally, we may turn this restrictive presentation of the S ratio round into a positive form and assert that any NR/OR ratio which is derived from two functionings having identical carbon grades for the two states and 100% and 0% oxidative substitution for the NR and OR states respectively must yield a pure S value.

On p. 187 when we come to speak of S_2 ratios we can produce instances in illustration of this principle.

The family of carbon grade values. In discussing the purity of S ratios we looked upon carbon-grade values merely as an intrusive distortion calling for scrutiny lest those of the numerator and denominator should not be identical. Now we may extend our outlook and take in such carbon-grade ratios as instruments of value in tracing changes of carbon activation rate, in which pursuit we desire to obtain pure C/C ratios undistorted by oxidation-grade differences. The wide range of carbon grade is shown by the carbactivation curve in fig. 10 and an arbitrary scale of values in units of x is appended as a relative to the CO_2 scale of the (TR-OR) curve. For location of points on the $[O_2]$ axis of this curve we may use the symbols Z, N, E.P., A and M proposed in the first presentation of the figure (see p. 181).

Returning to the previous section we can define the C/C ratio involved in the evaluation of S as being C_N for the numerator where the state examined is that of tissue adjusted by a 48 hr. exposure to pure nitrogen, while for the denominator the grade must be that defined as $C_{E.P.}$, since the tissue giving the $OR_{E.P.}$ value has been adapted to the O_2 supply characterizing the E.P. locus.

Clearly then the locations of the two carbon grades on the curve are different, and might be expected to distort the S ratio. But the form of the carbon curve in the figure expresses the postulate that the grade of C is identical all through the region of 'low oxygen' from C_N on to $C_{E.P.}$, where the carbon grade first begins to rise by the accumulation of free O_2 in the tissues.

By this special postulate then the two carbon grades are numerically equal and there is no distortion of the S value since $C_N/C_{E.P.} = 1\cdot0$. This minimal carbon grade extending from C_N to $C_{E.P.}$ may be conveniently spoken of as the 'anaerobic' carbon grade.

In the region above E.P. on the contrary, where the tissues are described as aerated, the carbon grade rises in a regular curve by oxygen acceleration of the nature outlined in description of the figure. It will be in this super E.P. region that variations of C grade will occur with variations of O_2 supply, and here we shall want to obtain pure C/C ratios as a compact expression of these changes. Absolute values of carbon grade in terms of metabolic carbon units are at present denied us, as this carbon metabolism covers the whole of the associated respiratory or fermentative metabolism and therefore has higher values of carbon than the measured absolute values of carbon escaping from the tissues as CO_2, which are our only index of magnitudes on the present scale of this analytic study of apple respiration. We can only deal in relative carbon grades between two loci. The anaerobic carbon grade would be a convenient standard of reference to which the higher grades such as C_A and ultimately C_M might be referred.

As it is postulated, in the next section, that oxidation grade is not a quantitatively distorting factor above E.P., it follows that observed OR values in this region should be a direct relative index of carbon grade. So our approach to the magnitudes of these pure ratios would simply be

$$\frac{OR_A}{OR_{E.P.}} = \frac{C_A}{C_{E.P.}} \quad \text{and} \quad \frac{C_M}{C_{E.P.}} = \frac{OR_M}{OR_{E.P.}}.$$

Such estimates of carbon grade need not be developed further here, but they will be of use in analysing mixed cases of decline of respiratory activity with senescence. And perhaps even more so in expounding the slow changes of carbon grade which form the long transitional phase leading to the adjusted state, always encountered when the external [O_2] of a tissue is suddenly changed so as to involve a new rate of carbactivation.

The sequence of oxidation grades. The general nature of this second major determinant of respiration rate has been expounded in connexion with the curve of these grades in fig. 10. This curve shows the grade rising with access of O_2 from zero in nitrogen to 100 % of the oxidizable carbon at the locus of E.P.; after which locus, by definition, there is excess of oxidizing agent, and the whole of the oxidizable carbon meets this fate however high the carbactivation rate is forced by the rising internal oxygen partial pressure.

The procedure for calculating the percentage grade arrived at for any $[O_2]$ between zero and E.P. is slightly complicated by the fact that the ratio of OR to NR in a mixed CO_2 output (TR) in low oxygen is not a measure of the ratio of respiration of carbon to fermentation of carbon in that metabolic state. The metabolic equivalent of unit OR in substitution is not unit NR but S units of NR. The S value for the tissue must be known before the oxidation grade can be computed.

A given carbactivation grade can yield either x units of OR or $x \times S$ units of NR. For any mixed state the oxidation percentage is best arrived at by the expression

$$\text{Oxidation grade (in \%)} = \frac{\text{OR} \times S}{\text{OR} \times S + \text{NR}} \times 100.$$

Oxidation grade has another character than the region of its control of CO_2 production which distinguishes it from carbon grade. While the adjustment of carbon grade on change of $[O_2]$ is extremely slow, presenting a long transition, the adjustment of oxidation grade is comparatively rapid. Time records of transitions show the two effects largely separated by their different rates of initiation and completion. This 'accident' of organization of the tissue catalyst provides the experimenter with opportunities of examining special transitional relations in which the carbon grade and oxidation grade show special couplings other than those which occur in normal fully adjusted state. Advantage is taken of these transitional couplings to obtain the ratios discussed in the next section.

A second evaluation of the substitution coefficient, S_2. If the difference of rate of adjustment for oxidation and carbactivation in apples were given an exaggerated magnitude, then we could imagine a state, after change from air to nitrogen, in which the carbon grade continues in exactly the state of being fully adjusted for air for say 12 hr. or more, though now in nitrogen. If, meanwhile, the oxidation grade had changed, rapidly and completely, to its adjusted 0% grade appropriate to nitrogen, then we should be examining a tissue which had a high carbon grade that we must label C_A, while the oxidation grade had become '0%', i.e. fully adjusted to nitrogen. This curious discord between the two determinants has been approached as far as actual differences of natural adjustment rate will permit by a block of experiments recorded in our earlier paper on nitrogen (Paper III). There we set out to measure what we called the ratio $\dfrac{\text{'initial NR'}}{\text{final OR}}$. For tissue long adjusted to air we measured the denominator value of OR and then immediately changed the atmosphere to pure nitrogen. A series of NR-max. values was recorded which declined fast through the O.G. transition from 100% grade to the adjusted 0% grade. The early observed NR-max. series was graphically extrapolated to the zero hour of entry into nitrogen, and this indicated a *still higher* '*initial* NR-max.' value than any of the observed NR-max. values. Had the graphic extrapolation no defects this initial NR-max. would be a CO_2 value appropriate to a combination of the high carbon grade C_A and the oxidation grade 0%.

What we actually measured then was an approximation to our theoretical complete discord of grades that we have just pictured. Such ratios then can be more fully defined in our present terminology as

$$\frac{\text{initial NR}}{\text{final OR}} = \frac{C_A \ [\text{O.G. } 0\%]}{C_A \ [\text{O.G. } 100\%]} = S_2.$$

The discord of oxygen relations shown by the two grades in the numerator cannot be maintained and actually examined in a normal tissue because adjustment of the C_A towards C_N must at once set in. Possibly by the technique of differential toxic inhibition of catalysts this state may be made available in the future.

The above expression for S_2 complies exactly with the criteria laid down on p. 184 as being essential for a ratio which measures the pure substitution coefficient, S.

The NR-max. value arrived at in this way is much above the NR_N value that starts our standard (TR-OR) curve in fig. 10. It had therefore been given a new suffix Z as a distorted form of the suffix N—distorted, we may say, by pre-history. We can therefore write the oxidation grades inside the bracket as $[NR_z]/[OR_A]$.

A table of the empirical ratios observed on application of this procedure will be found in Paper III (p. 71). There apples of class A gave S_2 value of about 1·5, while in our present study of S by use of low oxygen, values of 1·66 suggest themselves. Seeing how different has been the technique for arriving at these two types of evaluations of S, their close agreement, one with the other, seems to give good support to the postulates upon which the analysis is based.

Before leaving this section which introduces the suffix Z for this special initial state we must point out that Z is intended to be the index of a *class*. A whole series of NR_z values is indicated above the NR_N value in fig. 10. The highest NR_z would have the value of $OR_M \times S$, declining through $OR_A \times S$ down to $OR_{E.P} \times S$ which is NR_N. Had the apples been in 100 % oxygen before nitrogen then the initial state in nitrogen would have given still higher NR_z values; had they been in 10 % instead of 21 % O_2 then the initial NR_z value would have been lower. Generalizing this conception —NR_z— would always show a value which was proportional to the carbon grade that it has been preadjusted to. This proportionality coefficient should have the value of S in theory so that $NR_z = S \times OR$ of the antecedent carbon state.

Composite NR/OR *ratios.* When we propose to analyse a ratio which is made up of two values from the TR-OR curve which lie on opposite sides of the E.P divide, a good deal of accessory knowledge is required in order to clear up their double determination. For such values we may create a class of composite ratios to be labelled P.

An empirical ratio, which figured largely in the early Paper III, and which, at primitive stages of investigation, was the only ratio generally presented in the comparison of aerobic with anaerobic respiration, may be formulated as $\frac{\text{adapted NR-max.}}{\text{adapted } OR_A}$. In Paper III this composite ratio served as a general index of observed drift of NR and OR values. Though the oxidation grades of the numerator and denominator are respectively [0 %] and [100 %] as required for an evaluation of S, yet the carbon grades will be C_N and C_A respectively, and very far from being identical, so that the ratio will be quite distorted as a source of S.

Such a composite ratio could only be resolved by analysing it into a carbon-grade ratio and an oxidation-grade ratio. We could achieve this through $OR_{E.P.}$ by writing $\dfrac{\text{adapted NR-max.}}{OR_A} = \dfrac{\text{adapted NR-max.}}{\text{adapted } OR_{E.P.}} \times \dfrac{\text{adapted } OR_{E.P.}}{\text{adapted } OR_A}$. Of these two component ratios the first is exactly the standard S value, and would be known, while the second component is a ratio of OR values above the E.P. and therefore identical with the ratio of the carbon grades $C_{E.P.}/C_A$, as we have recently seen. Thus this composite ratio of the P type can then be analysed as being the product of two well-defined ratios of the S and the C class respectively. Each of these may receive support from other separate studies of these component ratios.

With this example we may close our survey of the classes of ratios that may be drawn from comparison of empirical CO_2 values between different parts of the (TR-OR) curve.

IX

THE INTERACTION OF FACTORS CONTROLLING RESPIRATION AND FERMENTATION IN THE HIGHER PLANTS*

CONTENTS

This paper is the ninth in a series of publications on the Problems of Plant Respiration† which have been encountered in the survey initiated some years ago in the Cambridge Botany School. It has seemed desirable at this stage to make a brief formal exposition of the position. Account is taken of the eight papers already published, and the way is also prepared for another set for which material has been collected, though preparation for publication can only be carried through slowly.

The present position is expounded as a formal array of the factors that have been found at work, and an attempt has been made to articulate these factors into a skeleton mechanism, which is sufficiently loose jointed to adapt itself to new positions with the progress of our inquiries.

The primary object in planning a systematic exploration of respiratory behaviour of the higher plants was that of accumulating well-grounded ideas about the vital directive control of respiration rate. Though respiration is so closely linked with

---

* This paper has not been published previously. The name F. F. Blackman appears on the title page of the typescript. Against the following heading for a final chapter—'Chap. IX, Comparison with the formulations of Meyerhof and Warburg for animal respiration'—is a note 'Not yet written. F.F.B. 27 July 1935'.

† Although this paper is the ninth in this book, Blackman had intended to publish it as Paper IX in his series 'Analytic Studies in Plant Respiration'; he was thinking here, therefore, both of the published series, which includes papers dealing with the respiration of the potato and of germinating barley seeds, and of the papers in this book.

our conception of vitality, yet its rate seems to be at the chance of a score of insignificant determinants.

Obscurity envelops any attempt to classify its significant aspects.

In accordance with this objective the first part is devoted to a survey of factorial controls.

After that, Part II will develop the second objective, which grew up in the early stages of the survey and centred in the problem of the metabolic relations of fermentation and respiration. This problem paraded itself before us because the earliest experimental studies aimed at seeing what happened when respiration rate was lowered by the reduction of oxygen supply. Here so-called anaerobic respiration was encountered, and the classical conception of Pfeffer upon its relation to aerobic respiration was found to be inadequate for the facts. Many facts that were quite unexpected came into view, and suggested that the effects of oxygen are multiple.

These aspects gained a heightened biological interest from the development, on the animal side also, of new views about the metabolic relations of fermentation and respiration.

The two types of plant tissue selected for our early surveys were those possessed of long vitality and also lacking the power of growth, such as apple fruits or isolated evergreen leaves. Only lately has the survey been extended to growing tissues, such as are provided by seedlings.

A wide survey of respiratory behaviour can only be carried through easily when it is limited to recording, as an index of respiration rate, the production of the end-product, CO_2, for the continuous estimation of which methods may readily be devised. Subsequently a technique for following also the intake of the reactant oxygen was worked out, and later still the survey was extended to the carbohydrate changes associated with respiration.

The condition that has been most widely studied, in all its variations, to bring out the varieties of respiratory behaviour has been that of the external oxygen supply. This has been explored over a range from zero to oxygen at 20 atmospheres' pressure. The results most drawn upon for the present survey concern the region of low oxygen concentration—say between 0 and 15 % O_2 in the surrounding air current. Here the facts revealed were strikingly at variance with current physiological teaching.

For physiological analysis we find it desirable to avoid the use of the phrase 'in air'. This carried an implication of 21 % O_2 outside the tissues while the internal aeration may be very various. We propose to use 'aerated state of the tissues'— as implying that *free* oxygen is present in the tissues and that all fermentation has been extinguished by oxidative substitution so that the aerated state is confined to the region along the (O_2) axis which is above the location of E.P.

PART I. SURVEY OF THE FACTORIAL CONTROLS REVEALED IN
ANALYTIC STUDIES OF RESPIRATION IN THE HIGHER PLANTS

The outlook upon respiration in this survey has been essentially kinetic. We have throughout pictured respiration as a reaction, but as a reaction taking place in a complex metabolic setting which involves also, within its ambit, both fermentation and carbohydrate metabolism. In the long enduring respiration which isolated parts of the higher plant can carry on, these other related metabolic processes have to be taken into account and the control of respiration becomes extrinsic rather than intrinsic. Even when we endeavour to confine ourselves to respiration in the narrowest sense we encounter facts that seem to prove that the essential nature of this reaction system is more complex than the presentations of it that have been accepted in the past.

The kinetic view of a metabolic process stresses, not only that it is a reaction which starts with the interaction of metabolic reactants and proceeds to the formation of determinable products, but also that these changes are under the control of a system of biocatalysts, the basic mechanism of which is of the type revealed by kinetic studies of enzyme reactions *in vitro*. The essential difference for reactions *in vivo* is that perpetuation of the reaction is provided for by other biocatalytic systems whose specific products become in their turn the reactants for the special reaction under examination. The perpetuated rate of respiration may thus be an exact index of the production rate of oxidizable carbon by some antecedent concurrent carbohydrate reaction.

The headings of the sections into which we propose to divide our survey will show our aim in making this formal presentation of the factors, though the facts were all gleaned in a rather disorderly manner upon extensive empirical exploration of respiratory phenomena.

1. *The factor of biocatalyst control of reaction rate*

Our early studies of respiration, when the only index of reaction rate used was the liberation of the product CO_2, showed that certain marked changes were spontaneously progressing in these living tissues, so that they did not maintain an identity of physiological state for long periods of time. As data for the respiration of leaves and of apples were accumulated this change of metabolic activity with development became very conspicuous. Later this progressive change appeared to be a universal phenomenon in higher plant development; and to characterize it we introduced the conception of an ontogenetic metabolic drift. Here we meet an internal cause of alterations of respiration rate from week to week that is quite inescapable. To deal with it empirically requires careful observation and adequate documentation. Attempting to interpret it we propose to segregate it with certain other factors under the general heading we have adopted for this chapter.

Among these other factors we may include as a second internal cause of variation of respiration rate, the physiological state of the tissues in relation to *starvation grade.* Our selected parts of plants, such as fruits, leaves or tubers, were often rich in reserve materials, typically polysaccharides, and ideally starch, which are slowly consumed during respiring existence in isolation. There may be enough reserves to maintain respiration for several months or even a year under laboratory or storage conditions of isolation. The most general manifestation of the starvation factor is a decline in respiration rate day by day.

A third cause of variation of respiration rate with constant external conditions arises from the slowness of adjustment of rate after the factor of external oxygen concentration has been suddenly altered. This is obviously of less wide application than the other two characteristics, and only has to be guarded against when tissues are transferred in the laboratory from their natural oxygen supply of 21 % in air to different artificial concentrations of O_2. This slow adjustment of rate to altered oxygen may last up to 48 hr., so that only after following a slow transition period of this length can the observer feel assured that the rate has become a true adjusted expression of the new oxygen concentration. With apples in particular, this type of behaviour is very marked.

We have now enumerated three characteristics of living plant tissues, the manifestations of which assert themselves in opposition to the experimenters' attempt to attain constant rates of respiration in return for provision of a constant make-up of the environment. Each one of these attributes must be recognized and allowed for before the precise grade of change of respiration rate, to be attributed to any such change as halving the external concentration of oxygen, can be arrived at. These attributes all have the appearance of spontaneous drifts of respiration rate, but they are drifts characterized by very different time relations. While the ontogeny drift is to be thought of in units of weeks, starvation may be said to alter in units of days and adjustment to oxygen concentration in units of hours.

The only practicable reaction of the investigator in presence of these 'spontaneous' drifts is to attempt to devise standards of reference for data that are influenced by them. With regard to the drift of 'adjustment' it is possible to determine its period and have knowledge of whether values are fully adjusted or semi-adjusted. As interesting secondary phenomena reveal themselves during the transition from one adjusted state to another, the delay of 48 hr. over adjustment is not without its informative scientific compensations for the investigator.

For standardizing the grade of starvation, when it is mixed up with lowering of respiration due to reduced oxygen supply, it is the practice to use the associated rate of respiration in air as the measure of starvation grade. To this air rate as a standard, the respiration rate in any other oxygen concentration can be referred, as a ratio. Respiration rates in air, such as are abundantly provided by controls, are symbolized by OR_A respiration values. The values found in low oxygen are symbolized as TR—total respiratory CO_2 values—this symbol having been adopted because the CO_2 may be partly the CO_2 of fermentation (NR) and partly CO_2 of aerobic respiration (OR). Adopting these definitions, the corrected rate of respiration determined by the low O_2 factor is arrived at by TR/OR_A, i.e. the ratio to respiration in air independent of concurrent starvation.

For correcting the ontogenetic drift effects no direct standardization is available, and it has to be accepted that values of TR/OR_A will not be identical at all stages of ontogeny. As many as four distinguishable graphic curves may be required to display fully the low oxygen relations of respiration to respiration in air for apples during their storage life in isolation.

With regard to these three classes of difficulties that appear as spontaneous changes in the biological properties of the tissues under examination, it is essential to ensure that the changes are not overlooked. Data can only be regarded as 'clean' respiration values when they can be certificated as (1) *time-adjusted* for O_2 changes, (2) *standardized to air for their starvation grade* and (3) *labelled fully for their ontogenetic phase*.

Turning now to possible interpretations of these biological factors, our survey has not been able to push critical consideration very far, but a few general aspects of the possibilities may be brought together here. We imagine that we are concerned with physiological controls that work through the organization of the protoplasm bearing the various biocatalytic colloidal systems. Such an 'organization control of reaction rate' may be considered to be an essential attribute of the living cell, balancing the fundamental accelerating effect of uncontrolled catalysts. About the precise changes that accompany such alterations of organization resistance we have only indirect knowledge, but study of enzyme kinetics *in vitro* provides us with models of processes that may be effective *in vivo*.

In vitro the changes of enzyme reaction rate are mostly associated with alterations of the effective amount of the catalyst E or of the enzyme substrate compound ES, but there are also, in complex systems, questions of diffusion rate of reactants in passage from one enzyme centre to another. In a reaction in an indifferent homogeneous medium such changes of diffusion resistance play no great part, but in a heterogeneous mobile medium, such as protoplasm, there seem to be possibilities of change which would produce marked variations in so-called protoplasmic permeability. These may be attended by striking, if ephemeral, physiological effects, typically of the nature of a rapid lowering of the resistance to reaction rate. If the physiological reaction showing this change of rate is under the control of the rate of production of its reactants by some distant catalytic centre, then after an ephemeral quickening the rate would return to the earlier controlled rate.

Passing on to more effective categories of control, we learn that *in vitro* the essential determiner of reaction rate is the amount of enzyme-substrate compound present at any moment. This can be lowered by the addition of depressant solutes or by inactivators of the amount of effective enzyme present. Depressant solutes are pictured as combining with E and so reducing the possibilities of ES combination. Two types of such depressants are recognized, the class which competes for E reversibly with the substrate and the class that is non-competitive in its union with E.

There is evidence available that respiration rate may be lowered by the generation of such a competitive depressant in the tissues of the potato upon exposure to low temperatures.

The activity of enzymes *in vitro* can in most cases be demonstrated to depend upon the maintenance of a complex particular system of which the enzymic centre is only a part, being in close association with carriers and co-bodies of various types. Removal of these other components may inactivate, as may also the proximity of other colloidal particles which adsorb, the enzyme system producing a more complex temporarily inactive system. From such a system the enzyme may be set free in its active form by appropriate eluents. These phenomena may well have their parallel *in vivo* and account for the rising and falling enzyme activity to which we may attribute ontogenetic metabolic drift.

Such drift might be general for all the catalysts in the protoplasm, or it might be differential as between such different systems as the oxidative and the fermentative. We have evidence that, in advanced senescence, the oxidative efficiency declines earlier than the fermentative system.

The investigator of these organization controls of metabolic rate is luckily not confined merely to observation and speculative interpretation; it is possible to carry out certain experimental alterations of organization.

Passing beyond the natural ontogenetic drift we have examined certain of these experimental conditions which, while not fatal, produce considerable changes in organization of protoplasm. Among these are wounding and mechanical stimulation. If an apple is cut into halves or quarters there is an enormous increase in CO_2 production—'the wound effect'. If the air round the apple is now replaced by nitrogen the increased CO_2 production is very much reduced, showing that there has been a differential effect as between F and R. On applying to a foliage leaf the stimulus of mechanical bending strains it is found that a marked rise of CO_2 production follows temporarily. This has yet to be examined in nitrogen. Other agents such as ethylene which rapidly advance senescent development (ripening of fruits) and increase respiration have also to be examined on metabolism in nitrogen.

Recently, quite a new chapter of research on specific catalytic control of metabolism has been opened by the study of drugs which have been discovered to be specific in their inhibiting power for certain biocatalysts. Some of these differentiate between respiratory enzymes and certain enzymes of the fermentation system. In particular, iodoacetates are held to inactivate fermentation without affecting respiration at all. Critical study of such effects is not at all easy with tissues of the higher plants, but our experimental work does not support such a simple picture as has been put forward.

2. *The mixed products of fermentation and respiration in low oxygen concentrations and their evaluation*

The part of our exploration of respiratory phenomena which bulks largest in the present exposition is the one concerned with respiration in low external oxygen concentration—'low $[O_2]$'. Placed in such artificial media, tissues may carry on simultaneously both fermentation metabolism, F, as well as respiratory metabolism, R. Each type may then be regarded as a possible metabolic factor in relation to the concurrent magnitude of the other type.

At this point we may stress that the symbol F stands for the whole fermentation metabolism as measured by its final products, expressed in carbon units, while R stands correspondingly for the total products, known and unknown, of aerobic respiration.

We have found it important to build up a formal approach to the facts which would enable us to distinguish F and R products and make out *the principles of their quantitative interaction*.

It is a complication in the study of such metabolic aspects in the higher plants that both F and R produce CO_2, while in animal metabolism lactic acid is the sole product of fermentation, CO_2 production being a clean index of respiration. Observing CO_2 production in any low $[O_2]$ we have to proceed to determine to what extent this mixed product TR is composed of respiratory OR and fermentative NR. TR values by themselves have no physiological significance until resolution has been achieved—at least approximately.

Before setting out upon this resolution of TR we must recall the lessons of the previous section that our observed values of TR will be distorted by the particular physiological state of the moment as regards (1) ontogeny, (2) starvation and (3) adaptation. If we wish to make wide use of the data we must have allowed due time to elapse for full adaptation, have corrected for the starvation grade and have labelled carefully for the ontogenetic phase. The correction for starvation is achieved as already pointed out by taking ratios to the TR values in air adopted as standard OR_A. We stress then once for all that every TR value used in this work is actually a TR/OR_A value.

Plotting our ordinates of TR/OR_A values for the range of oxygen concentrations between zero and the 21 % of air along an abscissa axis of $[O_2]$, certain approximate resolutions into components suggest themselves graphically, since NR must be maximal in nitrogen and OR be zero. As $[O_2]$ rises, OR rises with it and NR falls, and the form of some TR curves suggests that the decline of NR proceeds until NR is zero at some low concentration of O_2. At this point then the observed TR should be wholly OR. So we have suggested that at this concentration of O_2 is located an extinction point (E.P.) for NR. We picture then that with rising oxygen NR has been, step by step, substituted by OR.

Seeking the quantitative relations of this *substitution* of the CO_2 of F by the CO_2 of R we have to relate to rising OR, not the falling residual NR, which we observe, but the rising NR deficit which has been caused to disappear. We have pointed out in detail in Paper VII how the coefficient of this oxidative substitution, S, may be arrived at through the expressions

$$TR = OR + NR, \tag{1}$$

$$NR\text{-max.} = NR\text{-res.} + NR\text{-def.}, \tag{2}$$

$$\frac{NR\text{-def.}}{OR} = S. \tag{3}$$

There we have stressed also the different ways in which the evaluation of S may be approached. If S is constant in value with rising grades of substitution, then it is possible to resolve the observed TR curve into its components of NR and OR throughout the range of observation. When that is securely carried through we have no further use for the TR curve but shift our attention to the two curves, rising OR and falling NR, that we have thus substituted for it.

This procedure is fully set out in Papers VII and VIII. It had to be carried through first at a stage of investigation when the confirmation of measured oxygen intake was not yet available, but subsequent work reported in the next section puts this analysis of TR on a quite firm footing. This analysis provides us with material for a fundamental graphic presentation, one which we call the 'low [O_2] schema', namely, a graph setting out over an [O_2] abscissa axis NR and OR values between zero and air, and indicating the location of the extinction point. This schema will be amplified in § 3, when we shall be able also to bring to account in it the intake of O_2 and R.Q. values.

In this expanded form we have before us the fullest possible information about the gaseous reactants and products of F and R, namely, CO_2 and O_2. But this full exposition of 'gaseous exchange' turns out to be by no means a final goal but merely the threshold of a further adventure, for we are at once drawn into exploration of metabolism below the surface. For NR CO_2 is never the sole carbon product of fermentation, and OR CO_2 may often not be the sole carbon outcome of aerobic respiration.

We pass then to consider carbon products of another class, the possible non-volatile products linked up with the gaseous production of CO_2.

The situation in respect to non-volatile products is of course very different for the two concurrent processes of F and R. In most of the plants which produce CO_2 in the absence of oxygen there is also production of alcohol —AL. The fermentation here seems to be exactly of the type characteristic of yeast so that we can write $F = NR + AL$. Just as, when an increase of aerobic OR is brought about, we find the CO_2 of F dwindling, so also the alcohol production decreases *pari passu*.

What is extinguished, then, by rising R is not merely NR but the whole carbon material of F, and we are led to stress a new aspect of R, that of being the cause of F-def. values. From equation (3) $\dfrac{\text{NR-def.}}{\text{OR}} = S$ we pass over to

$$\frac{F\text{-def.} = (\text{NR-def.} + \text{AL-def.})}{\text{OR}},\tag{4}$$

which equals a larger value, say $S \times y$. For balance sheets of substitution we want to know the numerical value F/NR, here taken as y.

On the classical equation for alcoholic fermentation y equals 3·0 and, at present, we may be satisfied to adopt that ideal value and regard F (expressed in carbon units) as being in all cases equal to 3 NR. We shall then always write F-def. = 3 NR-def.

In only one case has a careful estimation been made of actual alcoholic production in relation to NR. We owe to Meirion Thomas and Fidler a special study of this aspect of metabolism in apples. They determine a value which they call 'the alcohol number', our y, which expresses the ratio $\dfrac{\text{NR} + \text{AL}}{\text{AL}}$. On the average y

comes out at about 2·57 for apples instead of the 3·0 of the simple fermentation equation. But until we get to final evaluations we will continue to use the value of $y = 3·0$, invariably, as connecting F values to the NR values, in carbon units.

Turning our attention to the products of aerobic respiration, the classic equation expresses this reaction solely as the complete oxidation of sugar to $CO_2 + H_2O$ with no other product in 'normal' cases.

In Paper III, however, we have taken up the position that associated with the catabolic oxidation there is a considerable further carbon metabolism of an oxidative anabolic type in which new carbohydrate is built up from lower-grade carbon products of fermentation. To this carbohydrate we may attach the symbol C_{OA} as a carbon compound formed by oxidative anabolism. We have then to ask what is the quantitative ratio between the carbon going to form C_{OA} and that oxidized to form the CO_2 of OR. If we use the symbol x for the relation of OA/OR, then we have the problem of estimating x where possible and also deciding whether it has an approximately constant value for similar physiological states. For apples we were led to assume that OA was large, being 3·5 to 4·0 times OR ($x = 3·5$ to 4·0). The line of thought that supported the hypothesis of such a large oxidative anabolism started from the observation that in oxidative substitution of OR for NR the total F-def. consists of about threefold the observed NR-def., so that to account for all this carbon in deficit some other fate must be available for it above the small value of OR that is the only demonstrable carbon arising in substitution. The aim of this outlook is to draw up a carbon balance sheet in which $C_{OA} + OR$ must together involve as many carbon units as are represented by the sum of NR deficit + AL deficit.

This picture elaborated independently many years ago has suggestive correspondence with the views initiated by Meyerhof for fermentation and respiration in muscle. To the fuller exposition of this correspondence we shall return in Part II.

Before leaving the category of substances that represent non-volatile carbon products of aerobic respiration we must make passing reference to the organic acids which are produced in large quantities in special cases. This type of respiration lies outside our present survey, but is also under investigation in an analytic study that had its origin in common with the present studies.

We may now pass from these earlier investigations of respiration by the output of *products* to the more limited range of cases in which quantitative studies of the consumption of *reactants* has been carried through.

3. *The oxygen reactant of respiration: oxygen intake and the gas exchange quotient*

After the respiratory behaviour of a number of types of plant tissues had been explored by recording their CO_2 production alone, the work then took up also the study of a reactant by measurements of oxygen intake. A special method was devised for getting long records of both aspects of gaseous exchange, and the first application of it was to the fermentation and respiration phenomena of apples in low concentrations of O_2.

When investigating CO_2 alone, our objective was to set out full data about products, over an abscissa axis of $[O_2]$; now we can aim at adding to the low $[O_2]$ schema various new values, the first of which is the observed O_2 intake, as it rises rapidly from zero in nitrogen. This constitutes the first category of new information, and we then open up the second derivative category which is to provide precise relations between O_2 intake and CO_2 production, occurring simultaneously. In very low O_2, the CO_2 production has the mixed nature of TR, but this can be analysed into its two components OR and NR. We could therefore set out three ratios of CO_2 values to O_2 intake, but here confine ourselves to $\dfrac{\text{TR}}{O_2\ \text{intake}}$ and $\dfrac{\text{OR}}{O_2\ \text{intake}}$.

An interesting situation develops out of the distinction between these two ratios. According to present custom both of them would be entitled 'respiratory quotients', but our closer analysis requires a definite distinction between the two. Historically, the term respiratory quotient was first used for relating CO_2/O_2 for respiration in air, where oxygen is in excess. It was held to provide an index of the nature or composition of the carbon compound serving as substrate for complete respiratory oxidation to CO_2 and H_2O. When, however, the external oxygen is low and the escaping TR contains appreciable fermentation products, then the ratio (no longer OR/O_2 but TR/O_2) is thereby falsified as an indicator of oxidation nature, and chiefly serves to indicate the relative contribution of F and R. We propose therefore to introduce the term *gas-exchange quotient*, shortened to 'gas-quotient', with the symbol G.Q. for the value of total observed CO_2, TR/O_2. The term R.Q. will be reserved for OR/O_2 when it is established that the numerator is really pure OR free from products of F.

The first tissue examined, that of the apple, presented us with marked complications between R.Q. and G.Q. values. Having worked out these values, they can be added to our low $[O_2]$ schema for each point where we have obtained data for O_2 intake and CO_2 production. G.Q. will obviously be infinite in nitrogen and fall at first very rapidly and then slower, reaching at the E.P. a low value which is maintained in spite of further oxygen increase. R.Q., on the other hand, should have the same value in all cases where there was any respiratory O_2 intake at all, remaining identical from observations made in air right back to those nearly in zero oxygen. By our definitions both R.Q. and G.Q. would each maintain an identical value in all regions of oxygen concentration above the E.P. In relation to this we have to report that ripe Bramley Seedling apples in a senescent state give values of $1\cdot2$ for TR/O_2 from about 10% oxygen right on to 30%.

If we had no additional facts to help us we might hesitate between several possible hypotheses, such as (1) that there is a fixed amount of the F reaction that persists unextinguishable by a big excess of oxygen and its aerobic respiration, or (2) that all F has been duly extinguished and that $1\cdot2$ is really a pure R.Q. quotient. This high R.Q. value might either be attributed to some reduction reaction (such as wax-formation from sugar) being coupled with the direct oxidation of sugar so that the intake of O_2 was reduced to subnormal by transference of O_2 from the reduction process to the oxidation process; or high R.Q. might be attributed to oxidation of malic acid to CO_2.

This last view is ruled out as an adequate explanation because analyses of acid loss show the amount disappearing is not nearly enough for this view.

The recent demonstration by Fidler that in those senescent stages of apples in which the G.Q. maintains the high value of 1·2 there is also a corresponding formation of alcohol enables us to decide in favour of our first hypothesis and adopt the view that the R.Q. in the narrow sense has the value of 1·00 owing to its representing purely an oxidation of carbohydrate.

These matters are more fully dealt with in the preceding paper (Watson).* It will there be seen clearly that ontogenetic drift brings marked changes in O_2 intakes with given external $[O_2]$, showing a shift with senescence in the direction of less oxidative efficiency. The pitch of G.Q. shows a marked rise with senescence, so that when the metabolism is failing the oxidation is not adequate to extinguish F, even in a gas of oxygen contents approaching normal air.

Considered formally these drifts are to be regarded as differential drifts between F and R metabolism and will come up again in that aspect in Part II.

4. *The carbon reactants of fermentation and respiration: the problems of specific carbohydrate substrates*

We have now to leave aside reactants and products which are in the easily identified form of gases and probe into the complications of the underlying non-volatile metabolites. We seek to characterize the nature of the carbon compounds which serve as reactants for fermentation and respiration, though indeed our aim is rather to travel hopefully along this road than to arrive.

For the present stage of our journey we exclude any question of the co-operation of fats or proteins, and persuade ourselves that F and R are concerned only with a group of metabolites for which we introduce the collective title of 'the carbohydrates and their relatives'

Though recent advances in the biochemical study of the mechanism of fermentation have made it possible to set out the sequence of metabolites for alcoholic fermentation, there is as yet no corresponding stability of view about the mechanism of aerobic oxidation.

Our own particular interest lies in the initial carbon reactants for these two processes, but, though there are to be found in the literature blocks of evidence supporting one specific sugar or another, as being the substrate for respiration, we cannot yet safely take advantage of this evidence to make satisfactory provisional pronouncements.

In all tissues that have been carefully studied, at least three sugars, the typical stable forms of sucrose, glucose and fructose are always present and we have no evidence that disqualifies, permanently, any one of them from being of primary significance in F or R. There are two ways in which the problem of correlating with R or F one out of a group of sugars, which we may call A, B, C, has been approached experimentally.

* See footnote p. 162.

The earliest and direct, simple approach was to try and determine by analysis which of the individual sugars disappeared during a period of functioning of R or of F. As two strictly comparable samples of tissue, as well as long periods of time have hitherto been required for production of such evidence, it is not a method of great applicability. We have several analytical studies in progress on the carbohydrates concerned in R and F but do not propose to present the indications of our data here.

There is one special complication in carbohydrate-metabolism that must be mentioned in this connexion, as thwarting the formulation of evidence on the lines we have just indicated. This concerns the abundant occurrence, within the group, of inter-conversions between the three grades of saccharides which may take place quite independently of F and R.

Such inter-conversions, though only of the nature of hydrolysis and condensation, have a definite linkage with the kinetics of R and F through concentration and equilibrium relations between inactive and active sugars. With isolated plant organs stored with polysaccharide reserves, hydrolysis is the greater producer of carbon compounds for the ultimate maintenance of F and R. So we may get this concurrent antecedent carbohydrate metabolism limiting the production rate of the inactive carbohydrates which precede active reactants, and so acting as the primary control of F and R rates in laboratory studies of such isolated tissues.

Modern biochemical research has shown also the possibility of spontaneous inter-conversion between the contrasted types of ring structure in sugar molecules, especially in the phosphorylated state.

We may now turn to the second line of approach to the problem of identifying, among a group of tissue sugars, $A + B + C$, which one it is that functions as a fundamental reactant to a function such as F or R. The principle upon which this approach is based is drawn from the field of enzyme kinetics where it has been established *in vitro* that when an enzyme, of amount E, is catalysing a substrate C to form the product P, then there will be a definite rate of production of P associated with each concentration of the substrate C. If rates of formation of P are plotted as ordinates against the abscissa axis of rising S then typically the rising values of P lie upon a curve which has the form of an enzymatic rectangular hyperbola, having its upper limiting value determined by E, the effective amount of enzyme.

If the respiratory system worked upon kinetic lines of a simplicity that would hardly be conceivable then it might be possible that some such simple picture would be applicable to respiration. In that case, when sugar A was being oxidized to OR one would expect to find a specific rate of R associated with each concentration of A, the whole set giving the relation of the enzymatic rectangular hyperbola.

We have tried to use this relation in an inverse way as a new instrument of research. Thus suppose there are three different sugars A, B, C in a respiring tissue, and that we are in a position to make analyses of their individual concentrations in association with a wide range of respiration rates. If one of the sugars only were the actual substrate of respiration, or closely related to it, then we should expect that sugar alone to give the predicated enzymatic relation to R.

This interrogation has been applied to respiring potatoes which can form sugars from the very large excess of starch in their tissues. From this source they produce the three sugars, sucrose, fructose, glucose. Of these it is found that sucrose alone shows the appropriate hyperbolic relation to the rate of respiration. We obtain thus a new line of *prima facie* evidence of a closer connexion of sucrose than of the hexoses to the mechanism of respiration.

This line of investigation involves a tedious number of analyses and its application has yet to be extended. It must be remembered too that though a very close correlation of sugar A with rate of function R may be established, this demonstration does not settle the nature of the mechanism of this connexion of A and R.

Instead of pursuing the details of evidence of this class it will perhaps be more in keeping with the type of presentation planned for this survey if we set out our problems formally using a special type of nomenclature. For the purpose of setting out the problems functionally we have prepared physiological labels for the different functions associated with F and R, which labels can be provisionally used in discussions while it is left to precise biochemical research to put forward molecules of definite types of structure as candidates for the differentiating labels.

Thus the carbon compound within the group—carbohydrates and their relatives —for which we ask identification as the starting point of fermentation will be labelled C_{NR}; while, correspondingly, the starting point for respiration will be labelled C_{OR}. It then becomes possible to discuss whether physiological evidence indicates that these two bodies are definitely separate substances or whether they are one and the same molecule.

Or, to speak by the book, we will ask whether the two labels should be attached to one molecule or to two separate ones. If the decision is for the latter position then it will be important to array the physiological evidence for the exact type of metabolic connexion that holds between them.

Besides C_{NR} and C_{OR} it will be important to complete the picture by introducing labels for the carbohydrates formed by synthetic anabolic processes from simpler carbon compounds. One of these stands out in this connexion and we have already adopted C_{OA} for the carbohydrate formed by oxidative anabolism in association with exothermic aerobic oxidation.

To this we may add C_{PA} as a symbol for the carbohydrate formed in photic anabolism. This would be the body, identification of which has often been sought for under the title of 'the first sugar of photosynthesis'. Recent studies of the respiration concurrent with assimilation in illuminated chlorophyll containing tissues make such a label as C_{PA} of schematic value.

In association with these problems of the higher plants, we might here recall the various lower organisms that produce their carbohydrates from CO_2 by what are called chemosynthetic mechanisms and note the problem as to whether the label C_{OA} might not be appropriate for these cases also.

The discussion of the possible relations of C_{OR}, C_{NR} and C_{OA} to the functions F and R will be taken up in detail in § 6 of Part II, see p. 203.

PART II. THE SPECIAL METABOLIC RELATIONS BETWEEN
RESPIRATION AND FERMENTATION IN THE HIGHER PLANTS

5. *The conception of oxidative anabolism* (OA), *and the fermentation deficit*

In this survey of respiration our first approach to the problems of the metabolic relations of respiration and fermentation was through studying in apples the changes in the rate of CO_2 production when the apple was passed as quickly as practicable from normal air to an atmosphere of pure nitrogen or reversely. Here we established the general proposition that the quantity of the change is such that to 1·00 OR there corresponds about 1·50 NR. Then again using quite a different setting of experimentation, we found on the average 1·66 NR being replaced by 1·00 OR. Within these numerical limits have fallen the results of scores of evaluations on substitution between F and R carried out on apples for a variety of physiological states and ontogenetic stages. One could only conclude that the recurrence of these ratios expressed a quantitative colligation of some sort between respiration and fermentation, and it seemed promising to regard the values as being part-expression of some change-over in a larger carbon metabolic balance sheet. Since, when the NR of F disappears, its inevitable associate AL disappears simultaneously the total carbon represented by the extinction of F must amount to about threefold NR. So we have a value of 4·5–5·0 carbon units lost, while only 1·0 carbon unit of OR is apparent on the aerobic side of the balance sheet. We are faced then with a fermentation deficit of about 4 carbon units for which some other fate has to be suggested. To explain this F deficit we introduced the conception of an oxidative anabolic function OA associated with the oxidation OR which results in the synthesis of a carbohydrate, provisionally labelled C_{OA} from intermediate reactants of F; OA having 3–4 times the magnitude of OR. Before this conception of a bulky aerobic synthesis, of which there is no outward sign, could be accepted as a general associate of OR, a number of indirect inquiries had to be set on foot about the relations of OA to the more obvious component reactions of F and R.

While these explorations were in progress, an association of anabolism and respiration was stressed by Meyerhof for the special metabolism of animal muscle in which, however, the fermentation product is entirely lactic acid, so that there is avoided the complication of the plant's dual production of CO_2. Later this view was extended by Meyerhof to yeast. A great impetus was given to this line of investigation for animal tissues by Warburg's discovery of differences in behaviour in these matters between normal and pathological tissues.

At the present time many workers are engaged in exploring various aspects of the relation of F and R for the whole range of animal tissues.

The central conception of the Meyerhof view is that the energy liberated by aerobic respiration in an animal tissue can be utilized in some way for building up the products of fermentation to carbohydrates. This reaction, labelled by Warburg the 'Pasteur reaction', evidently corresponds in many ways with the oxidative anabolism of plants. In a later section the two conceptions will be compared closely, in their qualitative and quantitative aspects.

For all this work on animal tissues, the living material is first cut into very thin slices and its gas exchange subsequently examined in closed vessels. This technique is unfortunately not applicable in general to the tissues of higher plants, for these are very quickly killed by such slicing. None of the tissues reported upon here can survive it, so that the approach of the plant physiologist is less direct. Recently a tissue with exceptional power of survival has been discovered and experiments with this material will be reported later.

6. *The problem of the carbon sources for F, R and* OA

In this section we shall turn our attention to the problems that envelop the carbon sources of respiration and fermentation metabolism, assuming, for our present purpose that these sources are limited to the carbohydrate family and its relatives. The fundamental question that we need ultimately to answer is: What exactly is the source of the carbon atoms that are perpetually being projected from plant cells, in combination with oxygen, as CO_2 of the type we call OR?

Widely different views have been adopted upon this subject, without much inquiry into the basis of the assumptions.

As a preliminary aid to our discussion of these matters in some detail we have prepared a set of labels, carrying only a functional significance, and ready to be attached (when knowledge permits it) to biochemically recognizable individual members, derivatives, or relatives of the sugar group. When this can be achieved we shall be able to follow their catabolic drift-lines of carbon through the welter of intermediate metabolism.

There are at present four labels C_{OR}, C_{NR}, C_{OA}, C_{PA}, each standing for a carbon-atom unit, but physiologically distinguished by their suffixes. Of these C_{NR} and C_{OR} are pictured as 'last' carbohydrates en route through catabolic breakdown for NR and OR respectively while C_{OA} and C_{PA} are pictured as 'first' carbohydrates arriving as new additions to the family as the products of anabolic processes. These four functional labels were more closely defined in the previous part, see p. 201, and we have now to search the metabolic evidence for any signs of their proper affiliation. For the easier presentation of this problem (which, borrowing nomenclature from Pirandello, we may entitle 'Four functions in search of their reactant') we have drawn up a schematic diagram, fig. 1 on p. 204. In this diagram the carbohydrates appear on the left, the end-products of respiration and fermentation to the right and the intermediate metabolism with various alternative progressions in the middle region.

The metabolism within the carbohydrate group is by no means confined to fermentation and respiration, so that we have been compelled to bring in at least three categories of carbohydrates involving three grades of complexity *A–B–C*, such as might be represented by polysaccharides, disaccharides, and monosaccharides. Causing interactions between these forms we have the bulky changes associated with hydrolysis and condensation. These changes constitute the *direct* interrelationships of carbohydrates, and they must not be ignored in stressing our special problem of the traffic lines of catabolism and anabolism which link up with carbon units outside the family. Unfortunately we are not yet assured of the mechanism of these direct reversible changes within the family any one of which if rapid might seriously distort the appearance of the special outside changes.

Interwoven with all this carbon metabolism is the problem of phosphorylation, a process which is steadily assuming larger and larger proportions as a contributor to the production of carbohydrate relatives of a much greater biochemical activity than the simple carbohydrates.

In spite of these clouds of obscurity that surround the metabolism of respiration on the carbohydrate side we are sufficiently encouraged to push on with an analytical formulation by the immense progress that has recently been made in the biochemical determination of the sequence of reactants involved in alcoholic fermentation of carbohydrates regarded as a self-contained process.

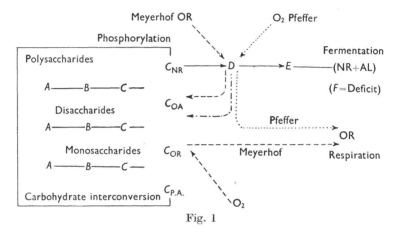

Fig. 1

For the development of our analysis we have drawn up the simplest schematic formulation of this fermentation sequence that will suffice to bring out the critical issues. In our diagram the fermentation sequence is represented by a linear series of carbon compounds A–F beginning with antecedent carbohydrate metabolism stages A–B–C and having C_{NR} as the last carbohydrate stage marking the initiation of the intermediate metabolism that leads on through stages D and E to the final products F which are identified biochemically as CO_2 plus alcohol (NR + AL). This alcoholic fermentation induced by the complete absence of oxygen is entirely outside the biological experience of the higher plants but we find that their power of maintaining pure F undiminished is very variable and their susceptibility to the toxic effect of the alcohol produced also shows wide differences. Only for the apple have we extensive direct studies of alcohol-production in nitrogen and later on we shall link up these studies which we owe to Meirion Thomas and Fidler to our own work on production rate of NR.

Perhaps the best approach to the problem of F which concerns us at the moment is to consider the exact effect of suddenly admitting oxygen and starting a change-over to respiratory metabolism. On doing this we find that in addition to the production of aerobic oxidation by the oxidation, to CO_2 and H_2O of some carbon compound assigned the label C_{OR}, there is also a diminution or complete extinction of the production of F (NR + AL). On this path of approach also there rises up before us the conception of a fermentation deficit which has for our type of quantitative analysis to be expressed in carbon units. This F-def. can only be arrived at indirectly by measuring first the maximal F-production in pure nitrogen and then the residual F values in the presence of a defined supply of oxygen. The F deficit in relation to this oxygen supply is then measured by F-max. less F-res. = F-def. When the whole of the F-max. is known to be extinguished then F-deficit must be equal in carbon units to F-max.

What the genetic relations of this F-def. may be to the concurrent R metabolism turns upon the essential nature of OR and the source of its carbon atoms, or expressed in terms of our presentation, the identity of the respiratory substrate C_{OR}. About this matter two very different views have been maintained. In animal physiology the view adopted from the earliest times was that the carbon compound oxidized was some single sugar which passed over to OR and H_2O by oxidation stages which were quite independent of any product that belongs to the fermentation line of reactants from C_{NR} on to F.

In the past history of physiology this view may be said to have been tacitly adopted as a simple hypothesis, rather than supported by experimental investigation. In recent times the contribution of Lundsgaard seemed at first decisive in favour of this view but later developments of his experimentation have shown that the evidence lacks the simplicity of application that was initially associated with it. Nor does our work on iodoacetate dosing of higher plants support such a simple interpretation for them either. We cannot therefore exclude from consideration an alternative view of the metabolism of R.

This second view, quite different from the first, was initiated by the plant-physiologist Pfeffer. Realizing fully the need for an explanation of the disappearance in air of the products of fermentation, he suggested that what respiration does is to interrupt the fermentation progress by oxidation of some of the intermediate links—say D or E on our diagram. If this oxidation is complete all the carbon of C_{NR} will be diverted to OR and no F products of NR + AL will arise. The label C_{OR} would then have to be attached to the F sequence somewhere about D or E.

Each of these two formulations has in modern times proved to be incomplete and each has been extended in the way that we must now consider.

With Pfeffer's theory the formulation seemed adequate, provided it was always found that the number of carbon units diverted into the form OR by oxidation of intermediates was just equal to the number that would (in the absence of oxidation) have gone on to appear as (NR + AL) in fermentation. In our own experimentation however we early met cases in which the carbon of this fermentation deficit is not thus equal to the carbon of OR but was indeed very much greater, so that some other fate than OR is needed for the larger part of the carbon in deficit.

To meet this difficulty we suggested that the fate of the rest of this carbon might be to be built back anabolically to some sugar-form which is to bear the label C_{OA} to indicate its origin by oxidative anabolism.

So far no plant has been observed which gives on oxidative substitution a greater carbon content in its OR than can have been derived from the carbon of its F-def. If such a one should be found then the mechanisms just enunciated will be judged as an inadequate proposal. In the diagram, the view of Pfeffer, as extended by our work on apples, is indicated by dotted tracks of genetic carbon drift. Here O_2 is shown as attacking the F chain at D and the resulting products are in part OR and in part C_{OA}. Here there is no independent C_{OR}.

We now turn back to the formulation of animal respiration which regarded the function as nothing but the oxidation of some sugar derivative C_{OR} to OR, and ignored at first the concurrent feature of arrest of fermentation. Later a new epoch was begun by the discovery of Meyerhof that the fermentation deficit in animal muscles has some quantitative relation of a general nature to the magnitude of aerobic respiration. Out of this developed the conception that there is a secondary, anabolic, reaction, called by Warburg the Pasteur reaction which is coupled with the energy liberated by the oxidation of C_{OR} to OR and that the anabolic product is a carbohydrate, formed from some fermentation stage such as D. We may thus conveniently apply the label C_{OA} to the product of the Meyerhof schema also.

In our diagram we have distinguished certain carbon lines of drift in this picture by broken lines. Here oxygen directly oxidizes the special carbohydrate C_{OR} giving rise only to OR as its carbon product. This OR production by its energy liberation is held to activate fermentation reactants, say again at D, and build back its carbon to the form of C_{OA}.

The practical distinctions between these two views of F and R metabolic relationships reside on the one hand in their propositions about the carbon sources and on the other hand in the fates of the F-def.

On the Meyerhofian view all the F-def. appears as C_{OA}; on the Pfefferian view it all appears as OR while on our extension of this view the F-def. is the source of both OR and C_{OA}.

For Meyerhof the carbon source of R, now labelled C_{OR}, is a carbohydrate independent of C_{NR} the source of fermentation and of C_{OA}. The sources of F and R are thus quite independent.

For the other view, in its strictest form, C_{NR} is the common source of both OR and C_{OA}; C_{OR} is reduced to the local status of an index of partition of carbon of the fermentation sequence between oxidation and anabolism.

Though, in order to economize words these views are expressed as antithetically as possible it is not intended here to give dogmatic support to the view that has been developed out of observations on plants. The intention of this article is rather to formulate the opposition of views as clearly as possible so that further data may be collected to clear up the critical issues between the theories.

For plants, such data are now in process of being brought to bear on the problems. One very general issue is an examination, for different physiological states, of rates of fermentation and rates of respiration in proper association so that it may be determined whether the two move always in sympathy, or whether F and R may be brought to show such independence of rate as to require the belief in the greater or less independence of the substrate, labelled C_{OR}, from that labelled C_{NR}. This inquiry should provide material for a new chapter to be entitled: What grade of independence exists between C_{OR} and C_{NR}?

7. *The evaluation of OA by measurement of substitution relations: complete and incomplete substitution*

We now formally adopt the conception that oxidative anabolism—OA— constitutes a second component of R (total respiratory metabolism, expressed in carbon units), and proceed to consider how the magnitude of OA can be measured and how the resulting values vary with ontogeny. In addition it will be shown that OA is a useful index of the exposition of the states of complete and incomplete oxidative substitution.

Since absolute magnitudes of OA vary, as do all our other magnitudes, it is essential to have a standard of reference for them and we may select its most objective associate namely OR and speak in terms of the ratio OA/OR—to be known as x. We may therefore regard x as a coefficient of anabolic efficiency of the oxidative function, OR.

In trying to make out the relation of OA to F and R and to oxidative substitution of carbon units we cannot measure x directly but can arrive at x values through the determination of two other quotients, namely $S = NR/TR$ and $G.Q. = TR/O_2$. Both these values S and $G.Q.$ are affected by x, and together they enable us to evaluate x in all conditions of oxidative substitution.

Complete oxidative substitution. Our approach to respiration is that of picturing a simple primary anaerobic state, F, in which all available carbon units are passing over to $NR + AL$ in the absence of oxygen. From this we change, on the admission of sufficient oxygen, to a state in which the carbon atoms of F have to a greater or less extent undergone oxidative substitution and are now appearing as $OR + OA$. The sum of $OR + OA$ constitutes the R metabolism. If substitution is complete the tissue has attained the 'aerated state' and the whole carbon metabolism has been redirected by oxygen, but still so that in carbon units $F = R$; however, besides this we shall have also to deal with cases in which oxidative substitution is only partial, so that there remains a residue of F still carrying on fermentation owing to the inadequacy of the oxidative machinery. In such cases we speak of a 'residual fermentation' converting a small number of carbon atoms to residual alcohol and residual CO_2: this we write as $f = al + nr$.

We may now consider drawing up our balance sheets of carbon metabolism. Seven cases are set out in Table 1. If we adopt 10 carbon atoms as the number to be dealt with in all substitutions, then, for complete substitution the process is $10\,F \rightarrow 10\,R$ with the substitution equivalence that $F = R = 10$. If, however, the whole of the $10\,F$ does not undergo substitution but leaves a few units, say $f = 2$, then the actual substitution is $(F - f) \rightarrow R = 8$ while the balance sheet of 10 units shows the categories on the two sides as $(F - f) + f = (R) + f$. The number of carbon atoms represented by f can be regarded as segregated from the rest and remaining unaltered through the oxidative substitution.

Analytic Studies in Plant Respiration

A heightened interest attaches to this type of analysis, which is to be set out in detail immediately, because there is a clear ontogenetic drift in apple development by which the oxidative functions expressed as OR and OA begin to fail before the fermentative, so that substitution during late senescence fails to be complete in aerated and fully aerated tissues and a definite amount of residual fermentation takes place in all concentrations of oxygen. If we make observations that are sharply defined enough to give trustworthy values of S, G.Q. and x we can, by an analysis, determine whether the oxidative failure is due to anabolic failure of OA alone or also to oxidative failure in the function OR.

Table 1. *Data for examples of oxidative substitution, arranged in sequence of columns, as used in carrying through analysis of the metabolism*

1	2	3	4	5	6	7	8	9	10	11	12	13
case	F	appa-rent S	NR	TR	G.Q.	intake O_2	OR	resid. NR	R	OA	x	correct S
A	10·0	1·66	3·33	2·00	1·00	2·00	2·00	0·00	10·0	8·0	4·0	$\dfrac{1\cdot66-0}{1\cdot00-0}=1\cdot66$
B		1·47		2·26	1·42	1·60	1·60	0·66	8·0	6·4	4·0	$\dfrac{3\cdot33-0\cdot66}{2\cdot26-0\cdot66}=1\cdot66$
C		1·32		2·52	2·10	1·20	1·20	1·32	6·0	4·8	4·0	$\dfrac{3\cdot33-1\cdot32}{2\cdot52-1\cdot32}=1\cdot66$
D		1·33		2·50	1·00	2·50	2·50	0·00	10·0	7·5	3·0	$\dfrac{1\cdot33-0}{1\cdot00-0}=1\cdot33$
E		1·25		2·66	1·33	2·00	2·00	0·66	8·0	6·0	3·0	$\dfrac{3\cdot33-0\cdot66}{2\cdot66-0\cdot66}=1\cdot33$
F		1·18		2·82	1·90	1·50	1·50	1·32	6·0	4·5	3·0	$\dfrac{3\cdot33-1\cdot32}{2\cdot82-1\cdot32}=1\cdot33$
G		1·00		3·32	1·66	2·00	2·00	1·32	6·0	4·0	2·0	$\dfrac{3\cdot33-1\cdot32}{3\cdot33-1\cdot33}=1\cdot00$

Cases A–D illustrate complete substitution; the others incomplete substitution. The set A–B–C shows decline of R efficiency with a constant value of $x=4\cdot0$. Set D–E–F shows a similar decline with $x=3\cdot0$. Sequence A–E–G illustrates decline of x values from $4\cdot0$ to $2\cdot0$, associated with constant oxidative OR.

We may now state the relations of the quotients to be used in our analysis of the complete or incomplete substitution between F and R. In the *complete* case the whole 10 units of carbon take part in substitution and our carbon balance-sheet can be written $(3\cdot33\,\text{NR}+6\cdot66\,\text{AL})=(2\cdot00\,\text{OR}+8\cdot00\,\text{OA})$. In this expression, two of our quotients, x and y, are explicitly stated. For the components of F we assume throughout the whole of this paper that the carbon appearing as NR is half that appearing as AL and therefore we can write always $F=3\,\text{NR}$ or $f=3\,nr$ as the case may be. The relation between OA and OR we define as x and this on the contrary we treat as having a certain range of variation, in the sense that x values tend to decline with development. In the present example we show in case A of the table x with the high value of $4\cdot0$. In addition to x there is implicit in our example the value of the oxidative substitution coefficient

$$S=\text{NR}/\text{TR}=3\cdot33/2\cdot00=1\cdot66.$$

Beside the coefficients x and S we need also to bring the intake of oxygen into the picture by determining the value of the gas exchange quotient G.Q. $=$ TR$/$O$_2$. Here the total CO$_2$ in air is all actually in the form of OR, so we find

$$\text{TR}/\text{O}_2 = 2\cdot00/2\cdot00 = 1\cdot00.$$

We must now note that x and S being fully defined in our formulation are strictly colligate so that knowledge of one gives us a value for the other. The relation is found as follows. If $S =$ NR$/$OR and $x =$ OA$/$OR, while $F = 3$ NR and $R =$ OA $+$ OR, then we get $F/$OR $= 3\,S$ and R$/$OR $= x + 1$.

As, on complete substitution $F = R$, we can then write

$$x = 3(S-1) \quad \text{and} \quad S = \tfrac{1}{3}(x+1).$$

Since S can be determined directly here we can thus calculate from the observations of case A that x has the value of 4·0. As an arbitrary example of a lower x value, we present case D, where $x = 3\cdot0$, and add the correlated change in S, should x decline to 3·0 while substitution remains complete. In this eventuality the oxidative side of the balance formulation would be 10 $R = (2\cdot5$ OR $+ 7\cdot5$ OA$)$. S would now be changed to 3·33 NR$/$2·5 OR $= 1\cdot33$ which exhibits the general numerical correlation stated above.

Incomplete substitution. The state of incomplete substitution which characterizes senescent apples in the aerated state requires a more detailed setting out. We may consider first the case where $f = 2$ after full admission of oxygen. There will then result 0·66 $nr = 1\cdot33\ al$ so that the total CO$_2$ given out in air (TR) will no longer be all OR but will give a higher G.Q. than 1·00; so demonstrating the incompleteness of the substitution. If x be still equal to 4 we must write

$$(3\cdot33\ \text{NR} + 6\cdot66\ \text{AL}) = [(1\cdot60\ \text{OR} + 6\cdot4\ \text{OA}) + (0\cdot66\ nr + 1\cdot33\ al)].$$

Contrasting this case, the data of which appear as line B in table 1, with the data for the case of complete substitution of line A, we note that R having fallen from an efficiency of 10 to the value of 8, both OR and OA, its components, are reduced, proportionately, maintaining thus the ratio of $x = 4\cdot0$; less oxygen intake falling from 2·00 to 1·60 now takes place, so maintaining the postulated equality of O$_2$ intake with true OR. The observed G.Q. $=$ TR$/$O$_2$ will in this case have risen to $\dfrac{1\cdot60 + 0\cdot66}{1\cdot60} = 1\cdot42$. This distortion by the presence of 0·66 nr will falsify the substitution coefficient also. Thus if we examine the substitution coefficient of such a case as B on passing it from air into nitrogen we get $S = \dfrac{\text{init. NR}}{\text{final TR}} = \dfrac{3\cdot33}{2\cdot26} = 1\cdot47$ instead of the $S = 1\cdot66$ of complete substitution.

Here then we must regard $S = 1\cdot47$ as an *apparent* substitution coefficient only, which needs correction because it is not really derived from a numerator and denominator which are in true substitution relations. If we want a true S value we must disentangle the residual nr from the gas exchange appearances. The simplest way is to subtract from both gas numerator and denominator the value of nr, as being outside the scope of the substitution. Hence the expression for correct substitution coefficient in case B is, as given in column 13,

$$\text{Corrected } S = \frac{\text{init. NR} - nr}{\text{TR} - nr} = \frac{3 \cdot 33 - 0 \cdot 66}{2 \cdot 26 - 0 \cdot 66} = \frac{2 \cdot 67}{1 \cdot 60} = 1 \cdot 66.$$

This corrected S value is naturally identical with that of typical complete substitution now that it is restricted to that part of the total gas exchange which undergoes complete substitution.

We may now glance at the effect of a further decline in the oxidative efficiency as illustrated by line C in table 1, where R has fallen to 6·0 units. As x is still kept at the value of 4·0, O_2 intake for OR will be reduced to 1·20 units. Now nr has risen to 1·32 units raising TR and lowering 'apparent S' to 1·32. The correct S, typical of perfect substitution, is again to be arrived at (see column 13) by subtracting residual nr from both NR and TR.

The cases A–B–C might represent an ontogenetic sequence through which a single apple would pass during storage. The rising G.Q. in the aerated states would indicate the general nature of the ontogenetic change and the successive determination of S would enable us to arrive at x and to decide that the oxidative decline was a *general* one affecting both OA and OR in equal proportion throughout the decline.

The block of data set out in lines D–E–F of table 1 show a similar *general* decline of R for the case where x has throughout the smaller value of 3·0. We note that the rising progressions of TR and G.Q. are here at a different increment. Such differences are shown also in the falling progression of S and O_2 through this series.

These two blocks of data for general decline of R, allow us to deal quickly with the type of case that we will call the *special* decline of R. Here it is supposed that OA is declining with ontogenetic drift but OR is remaining unaltered, thus providing a series where x is steadily declining, in contrast with the types just discussed above. The obvious character of this *special* decline is the striking one that oxygen-intake remains unaltered with decline of R though residual fermentation is steadily increasing. This series may be represented by the sequence A, E, G in the table. In this x is successively shown as 4·0, 3·0 and 2·0 while OR and O_2-intake remain at 1·0 all through the series. Here as O_2 is not declining G.Q. rises less steeply, 1·00, 1·33, 1·66, than for the *general* decline of R.

The data set out successively in the table have been drawn up to fit selected values of x and of nr, while from them the associated quotients G.Q. and S have been calculated. We think that the use of these formulations for the analysis of unknown metabolic states will best be made clear by working through the values of one example, case B, in the logical stages in which the arithmetical analysis is actually carried out. The columns in table 1 are arranged in the appropriate sequence for this purpose.

(1) We first make use of the observed 'apparent' S value which is, say, 1·47 on passing the apple from air to pure nitrogen. As the standard for our consideration is always 10 carbon units, the numerator, NR, will be 3·33 units for every case

in the present survey. Hence the derived denominator TR will be calculated from observed S as 2·26.

(2) TR being thus known we use its value as numerator in the observed G.Q. of 1·42 and arrive at the denominator 1·60 which gives the O_2-intake (in *carbon* units) which is by definition equal to OR. Independent check on O_2 values is given by direct estimation. .

(3) OR being known as 1·60 and TR as 2·26 we arrive at the value of nr by difference.

(4) nr being found to be 0·66 we allot f the threefold value of 2·0. Hence R only accounts for 8·0 carbon units.

(5) R being established as 8·0 and OR at 1·60, we get OA by difference as 6·4 units. Hence $x = \text{OA/OR}$ has the value 4·0.

(6) Having made full use of the 'apparent S' value we have now to correct it to yield the 'true substitution coefficient' for the actual part substitution. The values are $\dfrac{3\cdot 33 - 0\cdot 66}{2\cdot 26 - 0\cdot 66} = 1\cdot 66$ which is the true S value always to be associated with $x = 4$. By these stages a full formal analysis of the metabolism may be made by measurement of O_2-intake and CO_2 production in air, and of CO_2 production in nitrogen in relation to them.

The correctness of all this analysis depends firstly upon the postulate that the carbon balance sheet is a legitimate approach to the problem of oxidative substitution and secondly upon our assumption that OA is an actual objective component of the metabolic products to be included in R. The basal features of the analysis involve the following three gas values and the application of the two derived quotients S and G.Q.

(1) CO_2 nitrogen = NR $\left.\vphantom{\begin{matrix}a\\b\end{matrix}}\right\} S.$
(2) CO_2 in aerated state = TR $\left.\vphantom{\begin{matrix}a\\b\end{matrix}}\right.$
(3) O_2-intake in air = O_2 $\left.\vphantom{\begin{matrix}a\\b\end{matrix}}\right\}$ G.Q.

8. *The six functions involved in respiration rate: the dual functioning of oxygen*

Our formulation of the F/R mechanism points to the existence of some six independent functions which by their varying quantitative interaction control the reaction-rate for the formation of products by the compounded mechanism. It will therefore be desirable to supplement the schema of molecular categories which we presented in § 2 by a second one which attempts to display the imagined collective mechanism of control and its six component controls. These six are shown in fig. 2 and it is very suggestive of the metabolic complexity that the natures of the six are so different. I is an oxidation reaction-velocity, II, carb-activation, is a shifting of equilibrium by oxygen concentration, III an amount of biocatalyst, IV an energy activation, V an autogenous breakdown reaction and VI a biological carbon-starvation grade.

We may now set down a series of comments on the supposed working principles of this schema. We suggest that there is a central pool of the reactant C_{NR} from which arise three types of reaction, set out as proceeding towards the right of the schema, and giving different end products. The pool of C_{NR} is shown as being continually fed from inactive carbon sources indicated on the left as C (lacking a suffix).

The functions that are of prime interest to us are those of oxygen and it seems impossible to escape the view that oxygen acts in a duality of ways which we present as functions I and II.

The classic conception of oxygen's function is that of an oxidizing agent and in our detailed scheme function I presents O_2 as a reactant oxidizing C_{OR} to give rise to OR. As the concentration of external oxygen is increased and the internal $[O_2]$ increases, so also the intake of O_2, the consumption of C_{OR} and the production

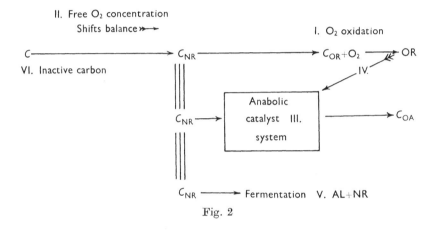

Fig. 2

of OR, all go on rising up to the external optimum of $[O_2]$, which may be as remote as pure 100% oxygen. This behaviour then has the appearance of an oxidation-function in which the rise in the maintained concentration of the reactant O_2 increases the rate of reaction.

But in addition to this simple oxidizing function I, we find evidence for another activity of oxygen in function II. We present oxygen as an activator of the rate of conversion of inactive C to the central reactant C_{NR}. Increase of this production-rate will also increase the production of C_{OR}, the carbon reactant for respiratory oxidation. Here then in contrast to function I we regard O_2 as increasing, by 'carbon activation', the production of oxidizable carbon for respiration.

We shall have presently to return to a detailed consideration of these two functions but this preliminary sketch of them will allow us to pass on to their four colleagues, functions III, IV, V and VI.

For function III we picture a system of anabolic catalysts which draws upon C_{NR} as its substrate and produces the anabolic product C_{OA}. If much energy of activation is needed to carry through this anabolic reaction then, in the absence of available radiant energy of light, some colligation with another exothermic reaction is supposed. In this case we have interpolated function IV as the energy-transference from the oxidation of C_{OR}.

The configuration of the schema implies that the rate of oxidation of C_{OR} would be a limiting control of rate of formation of C_{OA}. The amount of the anabolic catalytic system (function III) would set an upper limit to production of C_{OA}. When, as in senescence, OA appears to fail in relation to OR, then it is assumed that this is brought about by reduction of the amount of catalysts by enzymic inactivation. On this picture, OA could not survive OR or exceed some specific activation-relation to the energy of the process forming OR. Normally the inter-action of III and IV would determine the values of $x = OA/OR$ expounded in previous sections.

Function V represents the standard alcoholic fermentation of C_{NR}. This is the function which will account for the whole of the C_{NR} in the absence of oxygen; but now that we know that in late senescent apples this function may persist in air as a small residual fermentation—f—it is essential to include it in our aerobic mechanism. This residual fermentation has not yet been studied in detail. Its existence raises the question of how the mechanism provides for priority of reaction along lines I and III leaving a residue of C_{NR} for f.

The persistence of fermentation in air in senescent apples recalls the normal state of yeast cells whose fermentation persists in all concentrations of oxygen, the respiratory biocatalysts not being present in sufficient activity ever to substitute fermentation completely. Yeast then is never in the aerated state according to our definition on p. 190.

Dismissing these matters of the products arising from C_{NR} we may turn to function VI which stands for the conglomeration of processes that keep up the concentration of the inactive carbohydrates which we regard, biologically, as stored reserves to be drawn upon by activation for the basal metabolite of our system, C_{NR}. In most tissues these stores of inactive carbon exist in great excess over the daily draft made upon them by oxidation and fermentation reactions. It seems probable that constituents of the cell-walls are available to be drawn upon as well as the recognized special reserves.

With the activation to C_{NR} may well be associated phosphorylation in some way that will be revealed before long by the intensive work on this field of bio-chemistry, which has already yielded such brilliant results in fermentation studies.

We may now examine more closely the problem of the dual functioning of oxygen. The outstanding fact is that at all concentrations of O_2 except the lowest region it appears that, as external O_2 is increased, so, *pari passu*, not only does OR increase and the associated OA, but also there is increase of the 'potentiality' of fermentation, as shown by sudden application of nitrogen. This would indicate that the total $(F + R)$ metabolism, as measured in carbon units, must have been brought to a higher pitch by increase of external oxygen, through the range of the aerated states.

Recent studies of the late senescent stage of apples in which residual fermenta-tion—f—is taking place in air, show further the interesting fact that the absolute magnitude of this function also increases with rise of oxygen. The data show that this increase is directly proportional to the increase of the rate of production of OR and OA and the rate of increase of the intake of O_2. We are driven to consider that the rising oxygen concentration must be working through some link of the mechanism which is common to all the three end reactions. On our picture this must be the common pool of C_{NR}. Activation of production of N_{NR} from C should increase the reactions I, IV and V similarly. Hence our assignment of this particular nature to function II of oxygen.

Adopting this conception, we face the possibility that the raising of the rate of respiratory production of OR by $[O_2]$ might take place through the carbon reactant, indirectly, or through the O_2 reactant directly. We have to inquire which of these possesses the greater actuality. The answer appears to be that in the higher regions of oxygen supply, producing the aerated states, such as are natural to the life of plants, it is the action of function II which sends up respiration while in the very low external concentrations, such as are made available only artificially, in the laboratory, it is function I that is in control. Thinking of all this metabolism in terms of our 'low $[O_2]$ schema' which involves plotting functional activities over an abscissal axis of external oxygen concentration we decide that the distinction between the application of functions I and II is related to E.P. Function I controls in the low region below the critical E.P. value: function II controls rates in the aerated states above the E.P. value. In early papers this critical locus was stressed as the abscissal value for the 'extinction point' of fermentation and of its NR-production. We can still use the letters E.P. for the more fundamental significance that we now wish to stress, namely that this is also the 'exchange point' between the dual controls of I and II.*

We may now define the oxygen regions along the abscissa axis and examine the change over of controls more closely. Surveying the whole range of external $[O_2]$ we pass through an example of Sachs' conception of a minimum, an optimum and a maximum of O_2 for the respiratory function. There is zero production of OR in zero O_2, a highest value of OR at about 100% O_2 while in oxygen at a pressure of 20 atmospheres the OR-production is quickly brought down to zero, and the apple is killed. We do not yet know what function of O_2 it is that inactivates the protoplasm in this latter way, but we must conclude that beside the two accelerating functions I and II, there is a depressing function at work even below the optimum of 100% O_2. We have now to set out how in the sub-E.P. region acceleration is due to function I while in the super E.P. region, of aerated states, function II is in control.

We may now follow the progression of respiration rates towards the optimum, starting at zero oxygen, where the production rate of C_{NR} from C has a low steady value and the C_{NR} is going on to F products at the same rate as C_{NR} is being produced. On the first arrival in the apple cells of traces of O_2 following on a very low external concentration these traces will unite with some of the equilibrium concentration of C_{OR} that is maintained reversibly with C_{NR} in nitrogen. At first this rate of formation of OR by oxidation of C_{OR} will be much below the

---

* The E.P. of fermentation was introduced for that $[O_2]$ which raised OR to the level that so increased the F-deficit that the whole of F was extinguished. The present schema indicates that what actually happens is that oxidation of C_{OR} diverts more C_{NR} to replace it until that removal of C_{NR} in colligation with the production of C_{OA} equals the rate of production of C_{NR} from C. At this stage there is no longer detectable fermentation.

production rate of C_{OR} and all arriving O_2 will be consumed. With further small rise of O_2 supply from outside a still greater consumption of C_{OR} will be maintained; but still below the production rate, with no residue of free O_2 accumulating in the tissues. Let the next rise of O_2 supply bring us to the position that the arrival of O_2 through the diffusion resistance is just equal to the original rate of production of C_{OR}. This character should locate the exchange point where we arrive at the threshold of change over to control by function II. At this point, and through the aerated states with all higher arrivals of O_2, consumption of C_{OR} is 100 % of production of C_{OR}. The 'oxidation grade' of C_{OR} which has been steadily increasing through the 'sub-E.P.' phase now reaches 100 % and remains at that value.

The effect of further external $[O_2]$ is now to bring about a marked increase in the concentration of free O_2 in the tissues. It is this rising pressure of free O_2 which affects the mechanism of carbon activation; and the rate of production of C_{NR}, which is supposed to have been practically uniform through the whole sub-E.P. region, now starts to rise markedly. This rise continues with slackening activity, through the aerated states, as the internal $[O_2]$ rises right on through the air value up to the O_2 optimum at about 100 %. All through this super-E.P. region, OR is controlled by the rate of production of C_{OR} and not at all by the concentration of oxygen as a reactant in oxidation.

There are analogies for this type of mechanism in modern studies of intermediate metabolism, such as those in which the concentration of O_2 determines the distribution of glutathione between its oxidized and unoxidized forms, while one of these alone acts as catalyst regulator to a metabolic reaction rate.

If there is actually this assumed change in the oxygen control at the E.P., we should expect at that point some evidence of a break in the continuity of the curve which represents O_2-intake (or equally OR-production) when plotted over an oxygen-concentration axis.

Whether such a break is marked enough for convincing demonstration remains yet to be established. The type of experimentation required is rather elaborate and the material must be homogeneous. The expected break in the curve-form would be such that after a rectilinear rise of O_2-intake throughout the 'sub-E.P.' region there would take place a break away to a curve in the aerated super E.P. region.

Other curves which must show a marked break at the E.P. would be (1) the curve of $(F + R)$ metabolism in carbon units which should be a horizontal line from zero O_2 to the E.P. and then start on a sort of hyperbolic curve of the form of the O_2 intake curve, (2) the curve of percentage oxidation grade of C_{OR} which should rise from zero to 100 % in the sub-E.P. phase and then continue level at 100 %; and (3) the curve of G.Q. values, TR/O_2 which being near infinity in the lowest $[O_2]$ would fall to 1·0 at the E.P. in a mature apple or to a higher value (say 1·2) in a late senescent apple. In either case the G.Q. value registered at the E.P. remains unchanged with further rise of $[O_2]$ throughout the super-E.P. region. This type of curve has been clearly established for apples.*

Inquiries are in progress for testing more exactly the breaks in curve-form that have been taken as index of change in oxygen control.

* See footnote on p. 162.

In conclusion let us draw attention to the salient points that emerge from this survey. Speaking broadly these amount to a suggested triplication of the functions of oxygen in plant respiration. The two new ones that are to be added to the classic oxidation function will each provide a starting point for further experimental work to determine what grade of actuality and generality attaches to these conceptions. The two functions have been styled, on the one hand 'oxidative anabolism', on the other 'oxygen carbon-activation'.

Oxidative Anabolism. This conception developed out of the empirical observation of the ratio S known as the oxidative substitution coefficient. This related the magnitude of the escaping gas product of F with that of the gas product of R. In cases in which it can be established by the observation of the gas exchange quotient G.Q. that the substitution is complete (equality of F and of R being involved) then S can be used to get numerical valuations of the non-gaseous products left in the tissues. The situation conceived is sketched in the diagram (see fig. 3, p. 217) where the state of R in the aerated state appears above the horizontal line, while that in nitrogen is set out below the line. If the total carbon of F involved is equal to that of R (complete substitution) then the internal ratios of R and of F may be put as equal respectively to $(x+1)$ and (y) as has been discussed in § 2. Accepting all this, then S, being drawn from gas products only, must express only the divergence of the ratios between the gaseous and non-gaseous products for the anaerobic and aerobic states.

The logical completion of this conception introduces the respiratory product OA and determines its precise magnitude. By following up S we were led to suggest a new function for oxygen.

Instead of using in the text the expression $R/OR = (x+1)$ we have preferred the variant $OA/OR = x$ which brings in directly the oxidative anabolic function.

It is of some interest to indicate the range of magnitudes for OA and x that our observations have met with so far. It will be seen from fig. 3 that we can only get precise values of x when we have decided upon a value for y. Taking $y = 3$, as throughout the text, x will be equal to $3(S-1)$ and we set out a few calculated values in parallel lines:

$$
\begin{array}{ccccccc}
S = 1\cdot66 & 1\cdot33 & 1\cdot0 & 0\cdot66 & 0\cdot5 & 0\cdot33 \\
x = 4\cdot0 & 3\cdot0 & 2\cdot0 & 1\cdot0 & 0\cdot5 & 0\cdot00
\end{array}
$$

The highest S recorded for apples in the work here surveyed was $S = 1\cdot66$, giving x the value of $4\cdot0$. Should OA be twice OR ($x = 2$) just as AL is twice NR then S will be $1\cdot00$ and as much CO_2 would be given out in nitrogen as at $OR_{E.P.}$ in the aerated state. In a plant with the CO_2 production in nitrogen only half that of $OR_{E.P.}$ ($S = 0\cdot5$), then OA would be also half OR. The series ends with the limiting case that OA is non-existent ($x = $ zero). In this case the CO_2 in nitrogen will be only one-third of that of $OR_{E.P.}$. We may conclude the series by admitting that should a case be discovered in which S was below $0\cdot33$, then some radical revision of the conception would be called for. All the above paragraphs deal with cases of complete substitution only, but in § 8 we encountered the workings of incomplete substitution. There we showed how determination of G.Q. could be brought in as diagnostic for the grade of incompleteness when F was greater than R so that residual f persisted in air. In this way correction could be made and true values of S, and so of x, still be arrived at.

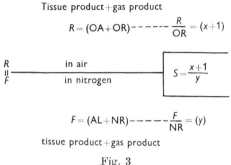

Tissue product + gas product

$$R = (OA + OR) - - - - - - \frac{R}{OR} = (x+1)$$

$$\begin{matrix} R \\ \| \\ F \end{matrix} \quad \begin{matrix} \text{in air} \\ \hline \text{in nitrogen} \end{matrix} \quad S = \frac{x+1}{y}$$

$$F = (AL + NR) - - - - - \frac{F}{NR} = (y)$$

tissue product + gas product

Fig. 3

Carbon-activation function of oxygen. The conception of an anabolic function for oxygen arose from the study of substitution between the F and the R metabolism. Combining these functions inside a square bracket expressing their sum $-[F + R]$, we may regard those studies as having been internal studies of respiratory metabolism, concerned with the make-up of the constituents of the total 'active carbon potentiality' which·the bracket represents.

Another line of study, which really runs concurrently with the internal study is concerned only with variations of the total absolute active carbon magnitude of the expression. This variation bulks largely in the life of the plant, marking the difference between normal and starvation states. Such variations we associate broadly with questions of sugar supply, and so far we have tried to regard these variations as a distortion and to eliminate them from our oxygen studies by standardizing our data relative to starvation grades.

Sugar effects will provide material for future chapters; we have already been led to conclude that oxygen also plays an important part in determining the totality of active carbon. We have assigned to it a function of accelerating the rate of production of active carbon compounds from those which are inactive from the point of view of direct participation in F and R metabolism. This function is discussed in § 8, where it is seen that the primary control of respirate rate by oxygen may prove to be of this nature.

APPENDIX I

Records of the primary data analysed in the nine papers

All the papers of this series of analytic studies deal with various aspects of the respiration phenomena observed in the prolonged investigation of a single set of apples from cool storage. For convenience of reference the primary data of the whole set are collected here as an appendix. As more than a thousand estimations of CO_2 production were carried out for this work the actual values are not printed in tables but are only presented as graphic records of the drift of respiration with time and various concentrations of oxygen.

This appendix contains, first, a table of characteristic attributes of each of the twenty-one apples; then some general notes on the graphic charts, followed by special notes on the individual records, and finally a series of charts in which are reproduced the graphic records of the respiration of the individual apples.

Notes to table. Each apple that was investigated was assigned a number and is identified by its roman numeral throughout these papers. These numbers run in chronological order from V to XXV, but as the respiration apparatus was a double one, two apples were generally under investigation at once. The dates in column 2 are a guide to this coupling, and often the two simultaneous records are presented in one chart. The colours of the apples at the date of column 2, when brought from cool storage to the respiration apparatus, are given in column 3 and serve as an index of their metabolic drift towards ripeness.

In the table the apples are not arranged in one chronological sequence but are segregated into three groups, representing the three physiological classes suggested in the first paper: class A ripens fast, class C very slowly, and class B at an intermediate rate.

Column 4 contains the fresh weight of the apple when the record starts. The duration of each observed record is given in hours in column 5. At the end of the record the apple was weighed again, and the loss recorded is expressed in column 6 as the loss per 10 days % of the initial weight. All respiration values are referred to 100 g. of *initial* weight. Correction of respiration values to the falling weight through the record has not been carried out; such formal correction would not seriously alter any of the significant values, and it is, further, open to objections on physiological grounds.

Owing to rise of temperature from 2·5 to 22° C. the initial course of CO_2 production has a complex rising form, and the true initial respiration value for 22° C. can only be got indirectly, by extrapolation back to zero hour of the record. The procedure for arriving at the theoretical initial values of CO_2 production is fully discussed in Paper I; the values finally adopted are set out in column 7, expressed as mg. CO_2 per 100 g. fresh weight per 3 hr. This value serves as starting point for the 'air-line' of respiration which figures conspicuously in the records. The values of air-line respiration 100 hr. later are given in column 13 and the drop in value appears in column 14, being the value of 13 less 7. The apples of class C are characterized by showing no fall of the air-line within 100 hr.

Table of characteristics of apples

1 Serial numbers	2 Date of record	3 Colour of apple	4 Initial weight	5 Duration of record	6 Loss of weight	7 Air-line initial	8 Slope initial	9 Hours to inflexion	10 Hours to maximum	11 Value at inflexion	12 Air-line type	13 Air value at 100 hr.	14 Fall in 100 hr.	15 Gas treatment (%)	16 Page for record
						Apples assigned to class A									
V	10. xi. 20	Green	174·2	210	0·77	11·75	13·6	32	10	11·6	II a	10·5	1·25	—	220
VI	10. xi. 20	Green	192·0	210	0·87	11·75	13·6	32	10	11·6	II a	10·5	1·25	—	220
IX	15. xii. 20	Green	126·5	140	1·32	14·0	15·6	25	10	13·9	II a	12·5	1·5	5	221
X	15. xii. 20	Green	144·0	140	1·73	14·0	15·6	25	10	13·9	II a	12·5	1·5	—	221
XI	11. i. 21	Green	127·4	170	1·57	16·8	18·1	20	10	16·8	II a	13·6	3·2	5, N	222
XII	11. i. 21	Less green	153·3	110	1·70	16·8	18·1	20	10	16·8	II a	15·1	1·7	—	221
XV	10. ii. 21	Green	156·8	240	—	20·5	—	—	—	—	II b	16·2	4·3	5	223
XVI	10. ii. 21	Green	183·0	240	—	19·5	—	—	—	—	II b	16·2	3·3	3	223
XVII	2. iii. 21	Green	132·0	220	1·52	20·6	22·2	22	9	19·6	II b	17·1	3·5	5, 7, 9	223
XIX	11. iv. 21	Yellow	162·6	150	1·84	16·2	19·0	38	11	14·4	III a	13·0	3·2	N	225
XX	30. iv. 21	Yellow	141·2	300	1·41	14·3	16·3	29	11	13·1	III a	11·2	3·1	5, 7, 9	227
XXI	30. iv. 21	Yellow	145·0	440	1·34	14·3	16·2	29	10	13·0	III a	11·1	3·2	N, N	227
						Apples assigned to class B									
XIII	24. i. 21	Green	139·5	200	1·60	14·2	15·4	16	8	13·5	II b	10·5	3·7	3, 5	222
XIV	24. i. 21	Green	149·0	200	2·09	14·6	16·5	27	10	13·5	II b	11·3	3·3	3	222
XVIII	2. iii. 21	Green	138·5	220	1·66	17·7	19·6	30	8	17·3	II b	16·3	1·3	—	225
XXIII	28. v. 21	Green yellow	126·1	240	1·67	17·6	19·1	19	13	16·7	III b	13·6	4·0	100, N	228
XXV	21. vi. 21	Golden	126·7	186	1·50	14·7	15·9	18	9	13·7	III b	11·1	3·6	N	229
						Apples assigned to class C									
VII	26. xi. 20	Green	134·0	320	0·99	12·8	14·25	24	9	12·9	I a	13·75		N	220
VIII	26. xi. 20	Green	147·0	300	1·01	12·2	13·5	23	8	12·6	I a	—	0	N	221
XXII	28. v. 21	Green yellow	145·8	256	2·03	17·3	18·6	23	13	17·6	I b	17·7	0	N	228
XXIV	21. vi. 21	Yellow green	138·3	230	1·81	18·5	20·1	29	11	18·8	I b	20·9	0	N	229

The initial CO_2 values set out in column 8 are the points at which a certain construction line would cut the ordinate axis at zero hour. This line gives the slope of the falling excess CO_2 production due to the initial heating up disturbance. A short length of this line is shown in each chart. Continued downward it would intersect the air-line at the 'inflexion' point, marked on the charts by a circle. The time in hours before this point is reached is set out in column 9; and the time to reach the actually observed maximum of CO_2 production on the way to the inflexion point is given in column 10. The CO_2 value at the inflexion point is given in column 11. This value constitutes the earliest observed respiration value on the

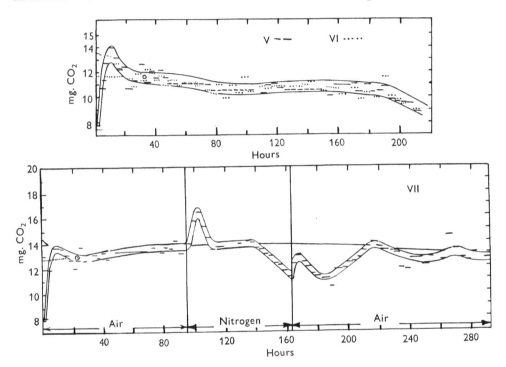

true air-line, and lies between the values in column 7 and column 13. The types of air-line drift are studied in Paper I and serve to characterize the three physiological classes. The entries in column 12 refer to the classification in Paper I, p. 12.

Lastly, in column 15 of the table are entered the various gas treatments other than air to which the apples were subjected. N indicates pure nitrogen, and the percentage numbers refer to oxygen concentration.

General notes on the charts. There are seventeen charts, containing in all twenty-two records, for the twenty-one apples numbered V–XXV. Apple XXI was observed for a very long period and its record is divided into two portions, XXI*a* and XXI*b*. All the charts are presented on a uniform scale as time drifts; the ordinate values indicate mg. CO_2 produced per 100 g. fresh initial weight of apple per 3 hr. The individual readings are all of 3 hr. duration and usually appear as heavy lines of this length in the charts. When two records of different individuals, observed simultaneously, appear in one chart, one of them is distinguished by a row of three heavy dots for each reading instead of a continuous line. These signs are also set against the roman numerals for identification of the two records in the one chart. The drifting series of readings which constitute the 'CO$_2$ record' right through the chart show nearly always the same amount of fluctuation about an imaginary mean line. Such a smooth mean line has not been drawn in the CO$_2$ records, but instead we have drawn a pair of 'contour lines' marking the upper and lower limits of the fluctuation of the CO$_2$ record. When an exact respiration

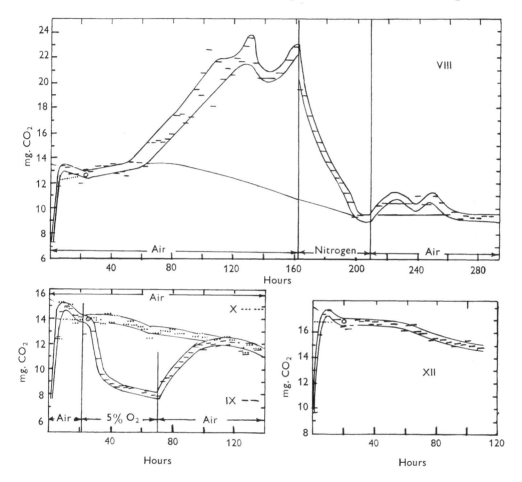

value is needed for our arguments the value midway between the contour lines has been adopted. There is obviously often some ambiguity about the location of the contour lines and the mean values; the precise turn of form that we have given to the records in various places is the outcome of a good deal of comparison and analysis. It is sometimes a matter of choice of interpretation rather than of proof of the relation put forward.

Every record starts with a longish period in air, and nearly all end with a period in air, so that the respiration drift in air may serve as a standard of reference. Only five of the apples remained in air all the time; the others were exposed to nitrogen or various concentrations of oxygen in the middle parts of the record. In column 15 of the table are set out their individual experiences, and in each chart there is a horizontal line indicating the sequence and duration of the various gas mixtures. In these gas mixtures the CO_2 production may depart widely from that in air, and the CO_2 record is drawn everywhere as a series of 3 hr. readings, enclosed in double contour lines; these contour lines are joined up with the air record so as to bring out the transitional course on passing from one gas to another.

When the apple is not in air the record contains an additional line, which is interpolated smoothly between the parts actually in air. This single line is in direct continuity with the imagined mean line between the contour lines in air.

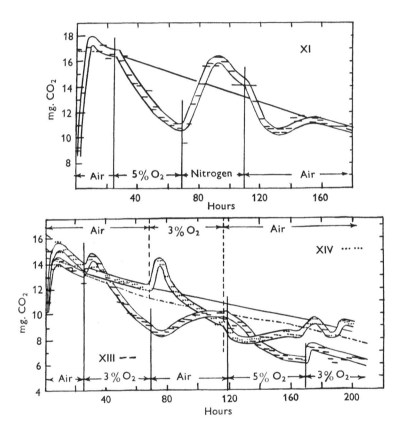

This track represents the course that the respiration of that particular apple would have followed had it been kept in air throughout the experiment. We thus get what we call the 'air-line' of respiration, ALR, right through the record, and this serves as an invaluable standard of reference in all these analytic studies. The use of this standard is discussed in the text at various places.

At the beginning of each record the course of respiration is much disturbed by the rapid rise of temperature from $2 \cdot 5°$ C. of cool storage to $22°$ C. of the respiration apparatus. Special study has been given to the 'initial rise of temperature effect' in the first paper of this series. Various construction symbols are inserted in the charts to characterize the magnitude of this effect. The circle between the contour lines locates the so-called inflexion point which marks the end of the initial disturbance and is thus the first observed point on the true air-line of the record. Carried back from this point is a broken line, extrapolated to cut the ordinate axis, and giving the theoretical beginning of the air-line. The value at zero hour gives the true *initial* respiration value (see column 7 of the table). Actually the CO_2 production has shot above this line temporarily and has then sloped down to the inflexion point. A short heavy line pointing down to the inflexion point is drawn

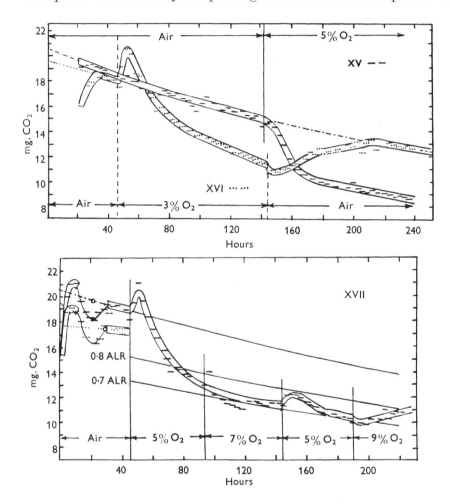

for a few hours from zero hour in each record, so that the initial value indicated by extrapolation of this slope is also brought out. The isolated heavy dot, some way beyond and below the circle, put in each chart serves merely to define more accurately the slope of this line through the inflexion point. The time duration of the phases of this initial temperature disturbance are set out in the columns of the table.

The symbols ALR, OR and NR, standing respectively for air-line respiration, oxygen respiration and nitrogen respiration, are defined more closely on p. 40 of Paper II.

Records V and VI (see p. 220). This was the first observation carried out and the two apples were kept in air throughout. In the chart the data for each apple are entered separately, V as lines and VI as sets of dots, but each record is not provided with its own pair of contour lines. Instead, two more widely separated contour lines enclose the double record, and the mean line between them would present the average air-line of the two apples. The form of the air-line beyond the inflexion point circle is well founded and drops very little until hour 180. It will be noted that scattered readings occur well outside the contour lines, but these are not taken any account of in our presentation of the records. The dotted line for the early part marks the ideal air-line, giving its initial and course, while the actual CO_2 record shoots temporarily up above it, giving the usual initial temperature effect. These two records are the type for early apples of class A.

Record VII (see p. 220). This record was continued for 320 hr. and even then showed but slight fall of the air-line. It is typical of early class C apples showing a rising air-line in the early part. The exposure to nitrogen gives the typical class C form, but this does not recur until apple XXII. In the middle of the nitrogen effect the nitrogen respiration values, NR, lie on the air-line. The track by which CO_2 returns in time to the ALR after nitrogen shows a striking characteristically transitional form.

Record VIII (see p. 221). This record was complicated by the active development of a patch of mycelium during the observation. It provides the only case of this observed. A full account of it is given in § 4 of Paper II. To the mycelium is due the rapid rise of respiration after hour 60. Nitrogen kills the mycelium, and the end of the record is believed to be pure apple respiration. This apple is assigned to class C. The continuous line for CO_2 production between hours 215 and 237 is due to stoppage of clockwork, so that seven 3 hr. readings were merged into one of 21 hr. The exact course of the record therefore becomes uncertain at this point, and the drift given to the contour lines here is based on the form observed in VII after nitrogen.

Records IX and X (see p. 221). These are two early apples of class A. Apple X was kept in air throughout the record, while IX was given 5% O_2 from hour 23 to hour 68. The air-line of X closely resembles that of V and VI, having the composite form which, after an initial level track, curves down from hour 40 to hour 70 and then starts another level track before curving down again. Record X here serves as a 'control' for the behaviour of IX in 5% O_2. We see the depression of respiration by this low oxygen down to very much below the air value of X. On return to air, IX rises in a slow smooth curve for about 45 hr. and then regains the value of the control, and proceeds afterwards along an identical track. We conclude that X represents also the course of the air-line of IX.

Record XI (see p. 222). This apple was treated with 5% O_2 after a rather short preliminary period in air, and its respiration is depressed in the same way as in record IX. After 44 hr. in 5% O_2 the apple was subjected to 39 hr. in nitrogen without any interval in air. This is the only case of such treatment, and the interpretation of the record presents several uncertainties. NR is clearly above the air-line when properly adjusted to nitrogen, and on returning to air the record dips far below the air-line and then climbs up to it again. These features are different from those of apple VII in nitrogen and mark this apple as belonging to class A. The form of drift of the air-line is not well established, as it requires an interpolation lasting about 130 hr. The clockwork stopped between hour 99 and hour 114, and a single reading, 15 hr. long, was alone available. No other early apple of class A was subjected to nitrogen, so we have nothing to compare this record with.

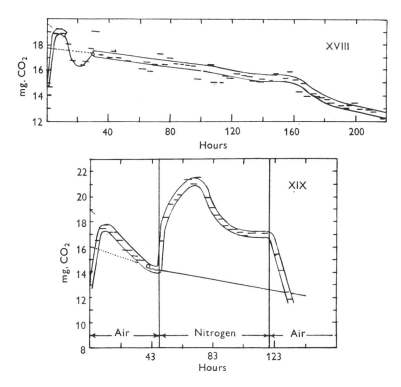

Record XII (see p. 221). This short record was in air throughout its course and gives an air-line of the type of V–VI–IX–X. It was carried out simultaneously with XI, but we have provisionally adopted a later type of air-line for XI.

Records XIII and XIV (see p. 222). These two class B apples were recorded simultaneously and subjected for periods to 3 and 5 % O_2. In the preliminary period in air XIV lies appreciably above XIII, and its initial temperature effect is shown as lasting to a later hour. Both low concentrations of oxygen depress the respiration, but 3 % give less low values of CO_2 production than 5 % relative to the air-line. The ultimate depression in 3 % is preceded by a preliminary rise of the CO_2 record above the air-line, an effect which is lacking in 5 %. These features are interpreted in detail in Paper IV. The recovery of XIV after 3 % follows a tortuous course, but ultimately it gives a location for the air-line, drawn as a continuous line. To XIII is assigned an air-line, drawn as a broken line, which is lower throughout and runs parallel to XIV. This location is supported by the values in air at hour 120, but the apple was not brought back to air at the end of its record.

Records XV and XVI (see p. 223). These two class A apples were recorded simultaneously. The first 30 hr. of the records were spoilt by an experimental failure, but the observed values from hour 30 to hour 48 have been extrapolated back to zero hour to locate the initial respiration values. Apple XV was kept on in air till hour 140, so that its air-line is extremely well established. After this it shows a typical depression of respiration by 5 % O_2, but it was not returned to air. Apple XVI was early given 3 % O_2 and shows exactly the same type of effect as records XIII and XIV. After 3 % it recovers in air, to give a track that is a perfect continuation of XV in air. It is therefore concluded that XV and XVI have identical air-line tracks after hour 40, before which XVI is slightly below XV. At this stage of senescent development the air-line approaches the type of a rectilinear decline throughout, thus differing from the early type of V and VI.

Record XVII (see p. 223). This apple, assigned to class A, was examined in 5, 7 and 9 % O_2 without return to air. The air-line drawn for it in the chart is derived from the common line of the two immediately preceding apples, XV and XVI. To facilitate evaluation of the grade of depression below the air-line which these various low concentrations of oxygen produce, two construction lines have been added, giving values for 0·8 ALR and 0·7 ALR respectively. The effect of 5 % O_2 on XVII differs from that of the early 5 % effects with IX and XI, and resembles rather that of early 3 %. This is an outcome of advancing senescent drift. At hour 12 the thermostat failed, and the drop of temperature is shown reaching a minimum at hour 20, after which it was readjusted and recovery takes place. The distorted readings in the chart are only connected by a mean line instead of by double contour lines. The early piece of record shown in sets of three dots up to hour 45 is that of XVIII, which was carried out simultaneously in air. This is left in this chart to demonstrate the identity of the effect of temperature distortion. There is a slighter temperature distortion between hours 108 and 120 when the temperature fell about 1° C., throwing the observed values below the contour lines. The dots in series above these displaced values represent temperature corrections to 22° C., by use of $Q_{10} = 2·5$.

Record XVIII (see p. 225). This apple is assigned to class B on consideration of its initial value of respiration and the form of drift of its air-line, which differs from adjacent class A apples. The apple was kept in air for 220 hr., so its air-line is perfectly established. It was carried out simultaneously with XVII and shows the same temperature distortions at hour 12 to hour 30 and hour 108 to hour 120 as we have just described.

Record XIX (see p. 225). In this record the effect of nitrogen was tried on an apple of class A, and we have a striking record showing how high NR rises above respiration in air. On return to air the CO_2 drops below the air-line before carrying out the characteristic recovery. In this record once more the heating failed at about hour 138. The later low readings up to hour 150 have been omitted from the record, as at this hour the experiment was discontinued and the air-line lost. The air-line here has no direct support in its later region, but with apples in this advanced stage of metabolic drift the form is well established by the next two records. The initial temperature effect is exceptionally marked here.

Record XX (see p. 227). This also is a late apple of class A and was used, like XVII, for investigation of 5, 7 and 9% O_2. Its air-line is well established by a preliminary 100 hr. in air, and a final return to air for 60 hr. at the end. It shows the typical smooth falling curve, getting less and less steep. Construction lines for 0·9, 0·8 and 0·7 of the air-line values are added to facilitate estimation of the depression produced by low percentages of oxygen.

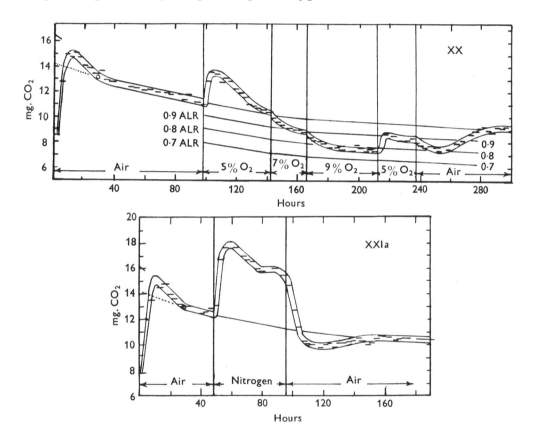

Record XXIa (see p. 227). This is the type apple of the late stages of class A. Its air-line is very well established with long periods of air. Nitrogen gives the typical effect, with NR high above ALR, and afterwards in air the CO_2 record dips below the air-line before recovering its position on it.

Record XXIb (see p. 228). This record is a direct continuation of XXIa on the same apple, but is separated from it because it contains a second exposure to nitrogen, given after complete recovery from the first. The two records together extend over 400 hr. and show how complete is the recovery from even 96 hr. in nitrogen. The air-line of XXIa is a falling curve slowly flattening out, while in XXIb this has passed over into a practically rectilinear slope of constant decline.

Record XXII (see p. 228). With this record we encounter again an apple of class C, which we have not met since VII, and the air-line drift rises slightly for a long time, so that after 100 hr. it is still above the initial value. Even at the end, about hour 260, the fall is very slight. The air-line is well established by 170 hr. in air at the beginning. When nitrogen is given the CO_2 record displays a form of exactly the type of VII, but not met, since then, with any of the nitrogen treatments. The recovery from nitrogen in air is also exactly like that of VII.

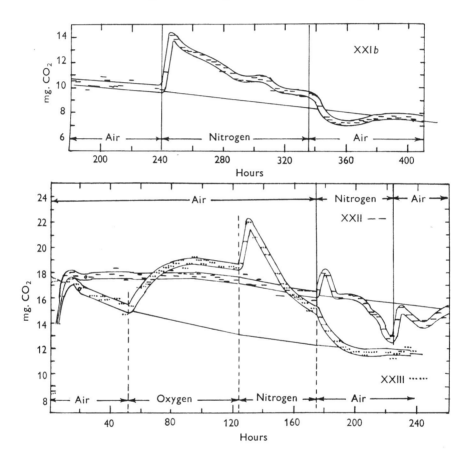

Record XXIII (see p. 228). This record was carried through simultaneously with XXII and appears in the same chart. The divergence of its air-line from that of XXII is very striking, seeing that they are so close initially. This apple exhibits the falling curve type of air-line, and is assigned to class B. This is the only case among the set of apples in which 100 % O_2 was given, and it exhibits the striking increase of CO_2 production, which rises gradually till it amounts to 1·4 times the air-line value. At hour 125 nitrogen was given directly after oxygen, leading to a further increase of CO_2 production, high above the air-line, as in the A–B type. At hour 174 the apple is returned to air and the CO_2 record falls to the air-line, dipping just a little below it about hour 200, before the subsequent rise of values about hour 230.

Record XXIV (see p. 229). Here we find the air-line rising during the first 100 hr., marking this out as a class C apple. The record in nitrogen starts with the form typical of class C, but the decline sets in quickly and steeply. Nitrogen was continued for 114 hr., and the continued fall and lack of adequate recovery in air demonstrates that a toxic effect has been brought about by nitrogen in this case.

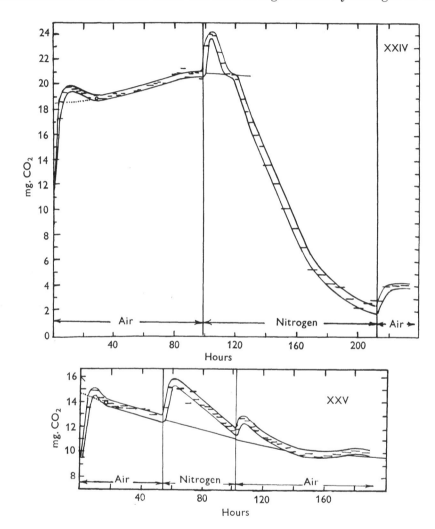

Record XXV (see p. 229). This record was carried through simultaneously with XXIV, but the form of the air-line is of the A–B type, like that of XXIII and unlike XXII and XXIV. The form of the record in nitrogen is also clearly of the A–B type, though the values fall off very fast. After nitrogen the record in air shows an initial rise of CO_2 production which does not occur in earlier apples of classes A and B. This we attribute to a specific after-effect of CO_2 production. When this is over, in about 45 hr., the air-line is again reached.

APPENDIX II

List of Names of Research Students of F. F. Blackman

Name	Title of dissertation submitted
AUDUS, L. J.	Further studies in the survival respiration of *Prunus lauro-cerasus*.
BARKER, J.	Certain aspects of the respiration of the tubers of the potato with particular reference to their significance in plant respiration.
BARNELL, H. R.	The respiration of barley seedlings.
BENNET-CLARK, T. A.	The metabolism of succulent plants.
BISHOP, L. R.	On the nitrogen metabolism of the barley plant.
BRIGGS, G. E.	
CALDWELL, J.	Studies in respiration. (*a*) The respiration of apples at pressures above atmospheric. (*b*) The respiration of leaves of various types.
CHAPMAN, R. E.	
CLAPHAM, A. R.	Sampling of agricultural crops.
COLLINS, E. J.	
DARWIN, N.	
DELF, E. M.	
DRUMMOND, J. M. F.	
EKAMBARAM, T.	The internal atmosphere and gaseous exchange of the respiring apple fruit.
FERGUSON, R. M.	Studies in respiration, with special reference to stimulation.
FORWARD, D. F.	The respiration of barley seedlings in relation to oxygen supply.
FOWERAKER, C. E.	
GIRTON, R. E.	
GODWIN, H.	The metabolism of starvation in leaves.
HANES, C. S.	Studies on plant amylases bearing on the problem of starch and sugar relationships in plant metabolism.
HAWKEY, S. P.	A study of injection in leaves and its effect on respiration.
HERKLOTS, G. A. C.	The effect of ethylene on the respiration of plant organs.
HULME, A. C.	Some biochemical studies in the nitrogen metabolism of the apple fruit.
INAMDAR, R. S.	
IRVING, A. A.	
JAMES, W. O.	On the relation of carbon dioxide concentration to the photosynthesis of submerged water plants.

Name	Title of dissertation submitted
KIDD, F.	
MASKELL, E. J.	The interaction of limiting factors in carbon assimilation with special reference to the 'CO$_2$ limited' leaf.
MATTHAEI, G. L. C.	
PARIJA, P.	
RAFIQUE AHMAD KHAN	Effect of various concentrations of oxygen on respiration of apples.
SHRI RANJAN	The influence of definite doses of sugar upon the respiration of leaves.
ROUX, E. R.	Plant respiration and temperature: the effect of temperature on respiration and carbohydrate balance in starving leaves.
SINGH, S. B.	Studies in the mechanism of respiration of massive plant tissues which offer great resistance to the passage of gases.
SMITH, A. J. M.	Studies in the storage environment of apples.
SMITH, A. M.	
SUMMERS, F.	
THODAY, D.	
TORRANCE, E. G.	
TURNER, J. S.	The relation between respiration and fermentation in excised carrot tissue, with special reference to the effect of sodium mono-iodoacetate on the metabolism of tissue slices.
VYVYAN, M. C.	
WATSON, D. J.	A study of the respiration of apples, with particular reference to oxygen consumption in respiration, and to the effect of change of oxygen concentration.
WILMOTT, A. J.	

Printed in the United States
By Bookmasters